Pearls from a Lost City

The Lvov School of Mathematics

HISTORY OF MATHEMATICS ❖ VOLUME 40

Pearls from a Lost City

The Lvov School of Mathematics

Roman Duda

Translated by
Daniel Davies

American Mathematical Society

This work was originally published in Polish by Wydawnictwo Uniwersytetu Wrocławskiego under the title "Lwowska Szkoła Matematyczna" ©2007. The present translation was created under license for the American Mathematical Society and is published by permission.

PUBLICATION SUBSIDIZED BY THE POLISH BOOK INSTITUTE

THE ©POLAND TRANSLATION PROGRAM

BOOK INSTITUTE ©POLAND

This publication has been subsidized by Instytut Książki—the ©POLAND Translation Programme.

A List of Illustrations and Acknowledgments is included at the end of the volume.

2010 *Mathematics Subject Classification.* Primary 01A60, 01A70, 01A72, 01A80.

For additional information and updates on this book, visit
www.ams.org/bookpages/hmath-40

Library of Congress Cataloging-in-Publication Data

Duda, Roman, author.
[Lwowska szkola matematyczna. English]
Pearls from a lost city : the Lvov school of mathematics / Roman Duda ; translated by Daniel Davies.
pages cm. — (History of mathematics ; volume 40)
Includes bibliographical references and index.
ISBN 978-1-4704-1076-6 (alk. paper)
1. Mathematics–Study and teaching–Ukraine–L′viv–History–20th century. 2. Mathematicians–Ukraine–L′viv–Biography. I. Title.
QA14.U382L85313 2014
510.71′24779—dc23

2013037212

∞ The paper used in this book is acid-free and falls within the guidelines
established to ensure permanence and durability.
Visit the AMS home page at `http://www.ams.org/`

10 9 8 7 6 5 4 3 2 1 19 18 17 16 15 14

Contents

Preface

When Poland gained its independence in 1918, having endured 123 years of servitude and oppression, the country was bloodied and ruined. For the next three years it would be further battered and torn by armed conflicts to its east and west (the Polish-Ukrainian War, the uprising in Poznań province, the Russo-Polish War against the Bolsheviks, the Silesian Uprising, as well as lesser uprisings). Once the borders had finally been settled in 1921 and peace arrived, nobody would have thought that "achieving independent status for Polish mathematics" (to quote Janiszewski) would happen so quickly in the wake of the political "strike for freedom" (to quote Kościuszko). But it did happen, and in an astonishing way. The phenomenon which the Polish School of Mathematics became, starting just after World War I, followed by its subsequent development up to World War II, has long been a source of awe and wonder. The school had two main branches, one in Warsaw, the other in Lvov. However, while much used to be written at home and abroad about the Warsaw school[1], there was total silence about the one in Lvov. The reason was political. In communist Poland it was impossible to write about anything linked, either directly or indirectly, with the territories seized by the Soviet Union: there was heavy censorship; a conspiracy of silence prevailed; even the city of Lvov itself was physically out of bounds, let alone getting anywhere near its archives. Thankfully those times are over, and one can now write freely about such things. It is now possible to travel to Lvov, and its archives are becoming ever more accessible. It therefore seems high time indeed for a reminder about the legacy and significance of the Lvov school. The issue is particularly pressing now that its founders and those who could still remember it have all passed away. At the same time, its "later grandsons" (to quote the poet Norvid) have a right to know what it was and what actually happened.

The aim of this book is to describe the Lvov School of Mathematics between 1920–1940. More precisely, the aim is to identify and describe: the time and place of true historical events; the key individuals and their followers; the school's greatest achievements and their significance. An attempt is also made to assess the influence of the school on mathematics in Poland as well as around the world. To do that, the book is divided into seven parts, each part subdivided as chapters. The first five parts describe the school itself. The last two provide complementary material. Specifically, the first part provides the background from which the school arose.

[1]We mention just a few of many sources: E. Szpilrajn (Marczewski), *Rozwój Matematyki w Polsce* (Development of mathematics in Poland), Cracow: Cracow Academy of Arts and Sciences, 1948; [B: Opiał, 1966], [B: Sierpiński, 1967]; [B: Kuratowski, 1989]; Sister Mary Grace Kuzawa, *Modern Mathematics: The Genesis of a School in Poland*, New Haven: New Haven College and University Press, 1968; [B: Kahane, 1995].

The second and third describe its development. The fourth part seeks to convey the drama of the tragedy which befell it, while the fifth is an attempt at understanding what happened and what its importance is. The sixth part provides short biograms of mathematicians who either belonged to the school or who where active there during its lifetime. The seventh part is an extensive bibliography, subdivided into three sections: Section A cites original works produced by Lvov mathematicians and quoted in the main text; Section B lists source material used by the author; while Section C references other mathematical works. For example, [A: Sierpiński, 1916b] refers to the second of Sierpiński's papers appearing in 1916 and is found in Section A, while [B: Steinhaus, 2010] refers to the personal recollections of H. Steinhaus, which were reprinted in 2010 and which can be found in Section B. Information about mathematicians from outside Lvov is given in accompanying footnotes.

As author, I wish to express my gratitude to Professor Roman Bobryk from Lvov (currently at the University of Liberal and Natural Sciences in Kielce) for making his archive material available to me, and to Professor Jarosław Prytuła of Lviv University for his kind support and help in numerous matters.

I also thank my reviewers, Professor Wojbor W. Woyczyński and Professor Andrzej Pelczar, for comments and favorable reviews, as well as others, including Andrzej Granas and Professor Lech Maligranda, who trial read the manuscript. A word of thanks is due to my editors, Irena Szymaniec and Lucyna Jachym, for their care and attention to the editorial layout, as well as for a favorable opinion.

* * *

Preface to the English edition

The first version of the book generated much interest in Poland, evidenced by eight favorable reviews in both specialist and general journals, the fact that it won the prize for Poland's best academic book of the year 2008, and some comments and updates received as feedback. Since the publication of the first edition new documents have come to light and some new works have appeared. This made it possible to implement minor corrections, provide extra substance to the main text in several places, while including some new entries in the bibliography.

The book has been generating significant interest abroad. In response to that interest, my earlier article [B: Duda, 2009] was translated into German and English, for which I received yet more feedback. The original text has been fully updated for the preparation of this volume, the first translation of the whole book into English. It is an honor to have the American Mathematical Society as publisher and for the book to be included in the AMS History of Mathematics Series. The translator is Mr. Daniel Davies, to whom I am most grateful for numerous incisive remarks, particularly regarding matters which might be familiar to Poles but are unlikely to be familiar to others. I owe a debt of gratitude to all concerned—reviewers, translators of my articles and books, and my readers. I particularly thank Professor Marek Górny for bringing the first edition to publication so wonderfully a few years ago and for encouraging me to update everything for this current issue.

* * *

Translator's notes and acknowledgements

Translating this work has been a challenge as well as an honor. However, it should be noted that this book is not intended exclusively for a specialist readership (though, admittedly, a mathematical education—at Imperial College and Warsaw University in my case—was certainly helpful for doing the actual translation). Indeed, the demonstrable success and interest generated by the original Polish version could not have happened were that the case. I therefore sincerely hope my translation will generate similar interest among English-speaking readers.

Bringing this work to publication was as much a moral imperative as a logical one. For far too many decades the subject matter of this award-winning book has been largely inaccessible to an English-speaking readership. Professor Duda provides us not only a long overdue, authoritative record of the achievements, significance, and personalities that made the Lvov School of Mathematics so special, he also profiles a very human story: a glimpse into the lost world of Lvov's once vibrant café culture, the complexity of underlying political and societal tensions, the need for an oppressed Polish community to assert both its cultural as well as its intellectual identity. Finally, there was the sheer brutality which preceded oblivion.

I would like to thank Dr. Sergei Gelfand (whose illustrious father is briefly mentioned in Chapter 22) and Dr. Ina Mette of the American Mathematical Society for approving this project, the AMS editorial team for their unwavering dedication, and the specialist reviewers for their erudite feedback. I am fully aware that even the best project ideas have no guarantee of making it to publication, especially now during these times of severe economic austerity. Most particularly, I must of course thank the author himself for confidently entrusting his precious text to my attention; having had the privilege of being invited to work for a few years at Wrocław University of Technology, I had the good fortune to be based in the author's home city, so I had the opportunity to discuss various matters personally with him. Lastly, I acknowledge several useful discussions with Mr. Marek Battek and Mr. Marek Łata regarding historical, cultural, and linguistic issues.

Any inaccuracies of translation are naturally my responsibility.

Daniel Davies
Oxford, United Kingdom

Part I. BACKGROUND

CHAPTER 1

The University and the Polytechnic in Lvov

Lvov University was the third oldest and the third ranked seat of learning to be established in the First Republic[2]. Similarly as in Vilnius, the capital of the Grand Duchy of Lithuania, it started off as a Jesuit college, one which had existed since 1608. King Jan Kazimierz transformed it in 1661, "conferring upon it the status of a university and the title of academy" [B: Finkel & Starzyński]. The Sejm, however, did not ratify the king's decree and the college remained a modest two-department academy (philosophy and theology) without the authority to award academic degrees. The act of foundation was passed by King August III in 1758, and one year later Pope Clement XIII issued a papal bull of approval. But the Cracow Academy (Akademia Krakowska) retaliated, using its power and influence to protect its monopoly in the Korona[3], and in 1764 the papal bull was revoked.

Starting in 1772, the Republic began being partitioned off, its lands systematically annexed and carved up between Austria, Prussia, and Russia. With it the Polish state was wiped off the political map of Europe for a period of 123 years (1795–1918). The occupying powers imposed a policy of assimilation on the territories they had seized, enforcing Germanization or Russification while suppressing the activities of Polish nationalists. The specific situation was different for each of the three seized territories.

Lvov became part of Austria in 1772 and stayed in its hands for almost 150 years, until the breakup of the Austro-Hungarian Empire in 1918. During this long period the land annexed by Austria was one of the largest Imperial provinces, known by the invaders as the Kingdom of Galicia and Lodomeria (or Galicia for short)[4].

The Academy in Lvov continued to function until 1773, when the Austrians closed it down after the abolition of the Jesuit order. They did not close it down for long, and for the next 100 years Lvov University managed to struggle on, but it never progressed much either. Modernized in 1784 by Emperor Joseph II into a four-department university with Latin as the language of curricular instruction, it lasted

[2]The one in Cracow was established in 1364, the one in Vilnius in 1578, and the one in Lvov in 1661. There were also smaller colleges, such as the Academy in Zamość, but which did not attain university status. From 1572 onwards, kings of the Polish-Lithuanian Commonwealth were elected and the role of Parliament grew stronger. For that reason the Commonwealth (up until 1795, when it ceased to exist) is customarily called the First Republic. Cf. N. Davies, *God's Playground. A History of Poland*, Vol I: *The origins to 1795*, Columbia Univ. Press, New York 1982.

[3]The First Republic actually consisted of two states. Sometimes known as the Two-Nation Republic, it formally consisted of the Kingdom of Poland and the Grand Duchy of Lithuania. The Korona (literally *The Crown*) refers to the old Polish part, i.e. Poland's ancient lands.

[4]A colorful description of Galicia between 1772–1918 is given in the book [B: Fras, 1999] and contains informative bibliographic references.

until 1805, when Emperor Francis I had it moved to Cracow. Revived once again in 1817 by Emperor Francis I, but now with German as the language of instruction, it remained a provincial Austrian seat of learning. In the nineteenth century Lvov emerged as one of the most important centers of Polish national culture. Galicia won regional autonomy in the 1860s, and its authorities (local governors, Sejm) were based thereafter in Lvov. In 1871 Polish was introduced as the language of curricular instruction. Its best period began in the years immediately preceding World War I, culminating in Lvov's twenty-year-long golden age of mathematics during the interwar years 1920–1940. Between 1919–1939 the university was known as Jan Kazimierz University (UJK for short), while from 1940 onwards it functioned as a Ukrainian university, named after the Ukrainian poet Ivan Frank.

The first mathematics department at Lvov University was set up in 1744. Over the course of the subsequent 200 years it would have many mathematics lecturers. In accordance with the prevailing tradition, universities concentrated on teaching (the cause of science being advanced through scientific societies and academies). Consequently university lectures tended to be at a rather elementary level. The National Education Commission (Komisja Edukacji Narodowej) implemented various reforms, including university reforms, during its active lifetime (1773–1994), but because Lvov found itself outside the state borders, it remained unaffected by the commission's reforms. After 1817, mathematics lectures were only given during the first year of study at the philosophy department, and for many years mathematics was at an elementary level. Kodesch was the first person to assume the mathematics chair. He was briefly succeeded by L. Schulz von Strasznitzki, a high-profile mathematician of the day, who was the first person to lecture higher level mathematics at Lvov (during the years 1834–1839). He was then succeeded by Ignacy Lemoch. All of them lectured in German. But the situation only really started to improve when Wawrzyniec Żmurko assumed the mathematics chair, a position he headed between 1871–1889. He also trained several mathematicians. The situation improved even further when Józef Puzyna took over in 1892. Puzyna is recognized in Poland as the first person in the world to write a monograph on analytic functions [A: Puzyna, 1898–1900][5]. Wacław Sierpiński obtained his habilitation degree in 1908 at Lvov University and was made an ordinary professor there in 1910[6]. Puzyna and Sierpiński were good mathematicians and mathematical life in Lvov benefitted accordingly, becoming much more dynamic (see Chapter 3). The fact that the theoretical physicist Marian Smoluchowski was at Lvov University (between 1900–1913) shows that it attained a rather high academic status in the years immediately preceding World War I. The lectures of those three people—Puzyna, Sierpiński, and Smoluchowski—were attended by Otton Nikodym, among others.

Besides the university, a technical school was established in Lvov during the nineteenth century. At first it was an intermediate-level school with a technical profile (1817–1825). In 1835 it became the Academy of Trade and Real Estate (Akademia Realno-Handlowa), and some of its students attended two-year courses. From 1844, when it became the Technical Academy (Akademia Techniczna), it

[5]See a description about it in the monograph [C: Saks & Zygmund, 1939] and in the article [B: Płoski, 1988].

[6]The decision of the philosophy department in 1913 to appoint W. Sierpiński to an ordinary professorship (he held an extraordinary professorship at the time) was blocked by the authorities. This may have been because of cost-cutting measures introduced by the Austro-Hungarian government, following the annexation of Bosnia and Hercegovina.

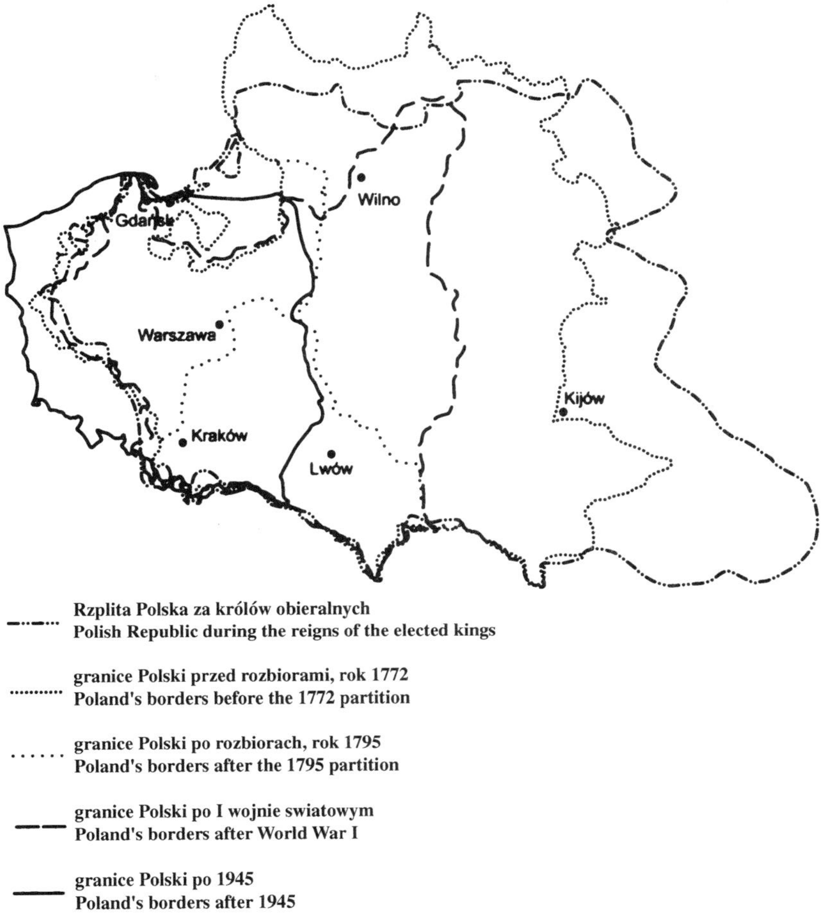

FIGURE 1. Lvov's changing borders 1772–1945. Established in the mid-thirteenth century, Lvov was a Polish city from the time of its acquisition in 1349 by King Kazimierz the Great until World War II. The war years 1939–1945 were the city's most tragic period. First the Soviets came and imposed Soviet order, which included the mass deportation of people. After the Soviets came the Germans, who cracked down heavily on the Polish intelligencia and committed genocide on the Jewish population. When they returned in 1944, the Soviets removed the entire remaining Polish population. Today Lvov is a totally different city—a Ukrainian city.

provided three-year courses, by which time it already had chairs in mathematics, physics, mechanics (which included descriptive geometry), and practical geometry (geodesy). After 1877, the academy (since changed to the Royal and Imperial Polytechnic School) was authorized to deliver lectures in the Polish language, but it was not until 1901 when it was given the right to award doctorates. In 1921 the polytechnic school became Lvov Polytechnic (Politechnika Lwowska).

The polytechnic school, just like the university, had some good mathematicians during the latter half of the nineteenth century and at the start of the twentieth century. Educated in the homeland, but completing their studies abroad, they published original mathematical papers and also wrote textbooks. They included Władysław Zajączkowski (ordinary professor during 1872–1898), Placyd Dziwiński (habilitation 1886, ordinary professor 1889–1928), Kazimierz Żorawski (habilitation 1892, private reader until 1895), Stanisław Kępiński (habilitation 1893, ordinary professor 1898–1908), and Zdzisław Krygowski (lecturer 1901, ordinary professor 1908–1919).

The last decades of the nineteenth century and the start of the twentieth century were a period of dynamic development in Lvov. As the city grew in importance, it grew in physical size too (this was the time when its most beautiful and most monumental buildings were constructed, such as the Galician Sejm, which later became the main building of UJK, and such as the theater, as well as others). With it, an interesting intellectual community emerged, revolving around the university, the polytechnic school, and the Ossolineum[7]. Kazimierz Twardowski's School of Philosophy gained ever greater status, physics was in a good state under Marian Smoluchowski, and mathematics was shaping up nicely for the foreseeable future (see Chapter 3).

[7]The Ossolineum is Poland's national archives institute. Originally founded in Lvov in 1817 from the private funds of an aristocrat, Count Ossoliński (1748–1834), it is now situated in the city of Wrocław, Poland.

CHAPTER 2

Polish Mathematics at the Turn of the Twentieth Century

In order to better understand later developments, let us familiarize ourselves with what Polish society looked like in the second half of the nineteenth century and what was the contemporary state of mathematics on Polish soil. After unsuccessful bids for independence, the last one being the January Uprising (1863–1864), the mindset of the country changed. Romantic acts of armed defiance were abandoned in favor of a less overtly confrontational approach, as was consistent with the mindset which characterized the prevailing age of positivism. Intellectual life was reawakened and mathematics too began to drag itself out of its relative obscurity during the first half of the nineteenth century. Ambitious young people went abroad to study mathematics, and some made a career there[8]. But others returned or studied in the homeland and contributed to intellectual life there, which was concentrated in three main places: Warsaw, Cracow, and Lvov. In Warsaw, after the closure of the Main School in 1869 and its replacement with the Imperial University, where lectures were delivered in Russian[9], Polish scholarly life principally grew outside of that university and was supported by the Mianowski Fund[10], Polish publishers[11], and by Polish scholarly organizations[12]. The university in Cracow continued to function, though lectures had been delivered in German for a very long time, as did the Cracow Scientific Society[13], which in 1872 became the Academy of Science and Arts (Akademia Umiejętności). The academy produced several

[8]Among them were J. Sochocki and J. Ptaszycki in St. Petersburg, W. Bortkiewicz in Berlin, and L. Lichtenstein in Leipzig, F. Mertens in Graz and Vienna (after two decades spent at the University in Cracow), A. B. Przeborski and C. Russjan in Kharkov, and J. Sleszyński in Odessa, H. G. Niewęgłowski and A. Sągajło in Paris, W. Folkierski and E. Habich in Lima, Peru.

[9]The University of Warsaw was established in 1816 but was closed down in 1832 in response to the November Uprising. It was restarted in 1862 under the guise of the Main School but was closed down again in 1869 in response to the January Uprising and was replaced with the Imperial University.

[10]In 1879 the idea was put forward for the creation of a Society for the Support of Polish Scholarship, but the refusal of the Russian authorities dictated it be called something else and the scope of its function be limited. Thus there appeared in 1881 the *Dr. Józef Mianowski (M.D.) assistance fund for persons working in scholarly fields*. For short, it was called the Mianowski Fund. It was in fact, and was deservedly called as such much later on, "the ministry for Polish scholarship during the years of oppression" ([B: Kasa, 1992]).

[11]For mathematics, the most important ones were "Prace Matematyczno-Fizyczne" (which appeared in the years 1887–1951) and "Wiadomości Matematyczne" (which appeared 1897–1939, with Series II running since 1955).

[12]The most important one was the Warsaw Scholarly Society (Towarzystwo Naukowe Warszawskie), established in 1907. It had the character of an academy, and it published "Reports from the sittings of the Towarzystwo Naukowe Warszawskie".

[13]It was actually the Towarzystwo Naukowe Krakowskie, which was founded in 1816. In the years 1817–1872 it produced "Yearbook of the Scholarly Society in coordination with the

journals, including the "Bulletin"[14]. Lvov too was a recognized center of culture, but we have already covered this (see Chapter 1).

These were humble but auspicious beginnings. The work of Polish authors in those days consisted of descriptions of results obtained by others or were lightweight, insignificant contributions[15], but some of them were referred to in the German review journal "Fortschritte"[16]. Mathematicians active in the homeland were already emerging and began to be recognized around the world.

> At the end of the nineteenth century, our mathematics could boast a series of undoubted achievements. The most important of these successes, on a countrywide scale, were some lasting journals that published mathematical papers, complemented by numerous and excellent libraries of basic textbooks on higher mathematics, the establishment of Polish mathematical terminology, there was a significant rise in the number of mathematicians compared with the paucity that had existed previously, and there was a gradual consolidation of scholarly milieux in Warsaw, Cracow, and Lvov. The papers of Mertens, Zaremba, Żorawski, and Sochocki, the mathematical devices of Żmurko and Abakanowicz, the ever greater accessibility of foreign journals to Polish authors, the strong links made with many leading mathematical centers in Germany, France, and Russia were the first successes on a European scale.
>
> [B: Opiał, 1966]

At the beginning of the twentieth century there was an identifiable, progressively upward trend in Polish mathematics. The endeavors of Warsaw in this regard were supported by the Mianowski Fund and by journals, above all by the journals Transactions in Mathematics and Physics (Prace Matematyczno-Fizyczne) and Mathematical News (Wiadomości Matematyczne). Moreover, it was there where there was, from 1907 onwards, an academy called the Warsaw Scholarly Society (Towarzystwo Naukowe Warszawskie), as well as a substitute Polish university called the Society of Scientific Courses (Towarzystwo Kursow Naukowych), a continuation of the illegal "flying" university (Latający Uniwersytet). The mathematics chairs in

University of Cracow", which after 1857 was changed to "Yearbook of the Royal and Imperial Scholarly Society of Cracow".

[14]Of importance to mathematics was the "Hearings and Reports of the sittings of the Department of Mathematics and Natural Sciences at the Academy of Arts and Sciences" (forty volumes appeared in 1874–1901). The "Bulletin International de l'Academie des Sciences de Cracovie", appearing since 1889, was the first Polish journal to publish mathematical papers in foreign languages.

[15]Four hundred fifty-nine Polish mathematical papers were written in the years 1873–1892, but they were only lightweight contributions, descriptions, reviews, etc.; almost all of them were written in Polish. See S. Dickstein, *Kilka słów o literaturze matematycznej polskiej w ciągu dwudziestolecia* 1873–1892, "Kosmos", 1895, pp. 352–358.

[16]In Berlin in 1868 there appeared the "Jahrbuch über die Fortschritte der Mathematik". It was the world's first ever review journal dedicated to mathematics. As of 1874, it included information prepared by S. Dickstein and M. Baraniecki about articles written by Polish mathematicians.

Cracow were assumed by Kazimierz Żorawski and Stanisław Zaremba[17], who were both by then reasonably well known around the world.

> With the appearance of these two scholars, Polish mathematics ceased being exclusively a consumer of ideas from abroad and of results obtained abroad. It was the moment when it started doing things by itself and engaged in the creative development of its subject. Conditions in those days were such that for a decade or two Stanisław Zaremba and Kazimierz Żorawski were the only representatives of Polish mathematics known outside their country.
>
> [B: Ślebodziński, 1969]

Lvov too was gaining ground, principally because of the efforts of Wacław Sierpiński (see the next chapter).

[17]Stanisław Zaremba (1863–1942). Born in the Ukraine, he finished his schooling in Petersburg, where he then studied (1881–1886) at the Technical Institute. He went to Paris after his studies, where he was awarded a doctorate in 1889. After working for a few years in French lycées, in 1900 he accepted an offer to work at the Jagiellonian University, where he became an ordinary professor in 1905. A member of the Academy of Science and Arts. An initiator and first president of the Polish Mathematical Society. Author of over 100 papers. His influence on mathematics in Cracow was substantial.

CHAPTER 3

Sierpiński's Stay at the University of Lvov (1908–1914)

Before World War I, the world's best scholarship stemmed from five countries (Germany, France, Great Britain, the USA, and Russia)[18], while the world's foremost capitals of mathematical excellence were considered to be Göttingen (where D. Hilbert, H. Minkowski, F. Klein, E. Landau, H. Weyl, C. Runge, and others were) and Paris (where H. Poincaré, C. Jordan, J. Hadamard, R. Baire, E. Borel, H. Lebesgue, and others were). Mathematical analysis was developing at a frantic pace; there was dynamic progress in algebra, while geometry was emerging from the crisis caused by the discovery of non-Euclidean geometry. But new areas of research were emerging, supported by some just as much as they were opposed by others. There was an emphasis on rigor and a tendency to express mathematics in the form of axiomatic-deductive systems. There was also a need to replace geometry, as the underlying basis for all mathematics, by a more abstract theory. This led to the arithmetization of analysis and the emergence of set theory, subsequently followed by the emergence of topology, so that set-theoretic methods became increasingly relevant in many diverse mathematical fields. D. Hilbert considered set theory "paradise"[19], while others were of the diametrically opposite persuasion[20]. Nobody, however, disputed that Cantor's set theory had opened up entirely new mathematical frontiers. There was but a single question to answer: should one take up the challenges of the new ideas or should one stick faithfully with the established, but still important, mainstream directions?

C. Jordan took up the challenge of the new, providing a long supplementary section in the second edition of his *Cours* [C: Jordan, 1893][21]. In it he framed fundamental notions of analysis in the language of set theory. There he formulated and proved (the proof was fallacious, due to a gap in the reasoning) his wonderful theorem on the cutting of the plane with a simple closed curve. Shortly afterwards, E. Borel presented the first version of the finite covering theorem [C: Borel, 1895],

[18]The concept of an International Catalogue of Mathematical Works arose towards the end of the nineteenth century (evidenced by a letter, dated 1899, of S. Dickstein to the Academy of Arts and Sciences in Cracow, regarding work on the catalogue). A Central Bureau was set up, to which the Germans sent 146,522 entries; France, 46,702; Great Britain, 43,484; the United States of America, 37,688; Russia, 21,071. Next on (that) list of entries are Poland, 3,492; India, Japan, and some other countries (Prace Matemtyczno-Fizyczne 9 (1905), Kronika, p. 121). Certain countries, notably Italy, did not respond to this initiative.

[19]"No one should be able to expel us from the paradise created for us by Cantor" (translated quotation in [B: Murawski, 1995; p. 126]).

[20]L. Kronecker opposed Cantorian set theory, while H. Poincaré raised certain fears too.

[21]A supplement to the book [C: Jordan, 1893], it is entitled *Note sur quelques points de la théorie des fonctions* and is found at the back of Volume II.

known today as the Borel-Lebesgue theorem, J. Hadamard proposed further research on continuous functions defined on the unit interval [0,1] in [C: Hadamard, 1898], while M. Fréchet gave a definition of metric spaces [C: Fréchet, 1906]. Thus began the modern theory of functions and set-theoretic topology. The work done in 1908 on the axiomatization of set theory ultimately led to it becoming a new foundation for mathematics itself.

These were the first, but important, manifestations of a new mathematics for the twentieth century.

New things were happening, and the great service done by the Lvov mathematicians was to recognize the new virgin territory and explore new research paths through it. The first contribution was the excellent, but unappreciated, monograph by J. Puzyna on analytic functions [A: Puzyna, 1898–1900]. It went unappreciated, because the academic environment in Poland was still not sufficiently advanced, and the fact that it was written in Polish limited its scope. The monograph introduced modern, clearly defined notions, based on set theory, to the theory of analytic functions. Its massive Chapter III, entitled *Z teoryi mnogości* ["from set theory"], presented some key topological notions, such as the "neighborhood" ["otoczenie"] of a point, a "closed set" ["mnogość zamknięta"], and a dense set ["mnogość wszędzie gęsta"], the continuum, etc.—as they were understood at the time. It was one of the very first set of lectures, anywhere in the world, to appear in book form on the subject of general topology (at least the first one after C. Jordan's *Cours*). Besides its influence on Lvov mathematics, Puzyna's monograph provided later inspiration for S. Saks and A. Zygmund [B: Kahane, 1995].

Józef Puzyna held a mathematics chair at the university in Lvov (from 1892 onwards). After the death of J. Rajewski, the ever dynamic Wacław Sierpiński obtained his habilitation there in 1908 and began giving lectures. As he recalled many years later [B: Sierpiński, 1967]:

> There had been no Polish mathematical congresses before World War I. There had only ever been mathematical sessions at congresses for Polish physicians[22] and natural scientists. In 1911 in Cracow, all four of the mathematics professors listed[23], as well as Dickstein[24] from Warsaw, attended just such a congress[25].

---

[22]This was the XI Congress of Polish Physicians and Natural Scientists (Cracow, July 18–22, 1911); see "Wiadomości Matematyczne" 15 (1911), pp. 289–291. The previous X Congress of Polish Physicians and Natural Scientists took place in Lvov (July 22–25, 1907); see "Wiadomości Matematyczne" 15 (1911).

[23]Kazimierz Żorawski, Stanisław Zaremba, Józef Puzyna, and Wacław Sierpiński.

[24]Samuel Dickstein (1851–1939), born in Warsaw, went to school there and between 1866–1870 he studied first at the Main School, then at the Imperial University. He taught in schools for many years, but above all he was very active in organizing academic life and dealing with publishers. He was a cofounder and coeditor of "Prace Matematyczno-Fizyczne" ("Transactions in mathematics and physics") and founder and chief editor of "Wiadomości Matematyczne" ("Mathematical News") . He wrote valuable articles on the history of mathematics and reviews of Polish mathematical papers, made numerous translations, popularized mathematics, attended mathematical congresses, worked with other centers, and corresponded frequently. In brief, he was a one-man institute all by himself. His services were recognized in the interwar Poland; see *Jubileusz 65-lecia dzialalności naukowej, pedagogicznej i społecznej Profesora Samuela Dicksteina* [Jubilee in recognition of Professor Samuel Dickstein's sixty-five years of academic service, teaching, and social activity], Warsaw 1932; see also "Wiadomości Matematyczne" 22.2 (1980), which contains some articles about Dickstein.

[25]In addition to those mentioned by Sierpiński, also attending the XI Congress were Zygmunt Janiszewski, Antoni Łomnicki, Alfred Rosenblatt, Edward Stamm, and Hugo Steinhaus.

> Each of us delivered their talks during a congressional session, but afterwards, during the informal meetings, we had friendly discussions about anything other than mathematics, because we were all working in different mathematical fields. [...]
>
> After that congress I came to the conclusion that such a state of affairs is not productive. At that time there was neither cooperation between mathematicians nor a system of mutual control. Although we did have mathematicians whose work made them known outside Poland, there was no such thing as Polish mathematics. I came to the conclusion it would be much better if a larger number of our mathematicians focused on a single field of research.

Sierpiński not only "came to the conclusion", he actually did something about it in Lvov, where he had been professor extraordinary at the university there since 1910. In the spring of 1910 he started giving regular lectures on set theory[26]; he was almost certainly the first person in the world to do so. His lectures encompassed general topology too, as evidenced by the book version of the lectures published a little later [A: Sierpiński, 1912a]. It was one of the world's first ever synthesis works of the theory. What is more, Sierpiński not only lectured, he also did research and drew younger people into the subject. From 1911 he gave a senior seminar on mathematics, which attracted quite a large group of younger members, active students, and researchers. Among them were Kazimierz Ajdukiewicz, Tadeusz Czeżowski, Włodzimierz Lewickij, Antoni Łomnicki, Otton Nikodym, Stanisław Ruziewicz, Miron Zaryckij, Zygmunt Zawirski, and others. It also attracted people from beyond. Zygmunt Janiszewski, who had already written a superb doctoral thesis on topology [A: Janiszewski, 1911], awarded in Paris, accepted W. Sierpiński's invitation to come to Lvov, where he obtained his habilitation in 1913. That same year Stefan Mazurkiewicz and Stanisław Ruziewicz obtained their doctorates at Lvov. They both had Józef Puzyna for their formal academic advisor, but Wacław Sierpiński was unquestionably the leader of the group, and his influence can be seen in the fact that four mathematicians (Sierpiński, Janiszewski, Mazurkiewicz, Ruziewicz) had very similar interests, revolving around set theory and its (as called at the time) "applications", i.e. the theory of functions and topology. The very first results obtained by the members of this group bear testimony to its productivity and originality.

W. Sierpiński was initially interested in number theory but switched his interests to set theory after coming to Lvov. He started working on what was then Cantor's astonishing theorem about the existence of a one-to-one mapping of the plane to the line, only he gave a different, elegant proof [A: Sierpiński, 1908]. And in this spirit, correcting some earlier results, he wrote other similar papers, such as [A: Sierpiński, 1910]. Around that time he also began publishing papers in topology, including one in which he constructed a square-filling curve [A: Sierpiński, 1912c]; see also [A: Sierpiński, 1912b]. That elegant geometric construction replaced the primitive analytic one [C: Peano, 1890] and is still widely cited today. Not long afterwards he stated his constructions of his now well-known triangular curve and his

[26]See the *Spring Lectures Program* at Francis Joseph University for 1909/1910, Lvov, 1910, p. 8.

"carpet" (stated in [A: Sierpiński, 1915], full constructions given in [A: Sierpiński, 1916c]).

To better understand Sierpiński's original constructions, we shall recall their underlying contextual background and their significance. Research into set-theoretic topology led to the discovery of more and more unexpected sets which were hard to classify. Yet at the same time greater precision was being demanded in reference to basic intuitive notions, including the intuitive understanding of what a curve actually is, even a curve lying in the two-dimensional plane (i.e. a definition restricted to plane curves). However, there were two competing propositions: a Jordan curve (a plane continuum which is the continuous image of an interval) and a Cantor curve (a plane continuum with empty interior). The difficulty was that both these definitions were counterintuitive. The square-filling curve ([C: Peano, 1890], the geometric construction was given in [A: Sierpiński, 1912b]), is a Jordan curve, whereas there exists a Cantor curve which does not contain a single arc [A: Janiszewski, 1912]. W. Sierpiński constructed two curves, a triangular curve and a "carpet". Although each of them is simultaneously a Jordan curve as well as a Cantor curve, they have paradoxical properties nonetheless. It meant that even if one could somehow merge both propositions into a single one, it would still not resolve the problem at hand. Sierpiński's curves left a permanent legacy in topology, inspiring the imagination as well as some later research. Furthermore, they forged new theoretical inroads. For example, Sierpiński's "carpet" was a universal Cantor curve; in other words, every Cantor curve can be embedded homeomorphically into the "carpet". So, Sierpiński's curve introduced to topology the notion of universal sets. As it later transpired, the two Sierpiński curves were also among the first ever examples of fractals.

We should add here that another reason why contemporary descriptions of a plane curve were unacceptable was because they were "external", in the sense that they made essential reference to something beyond themselves: Jordan curves were defined in reference to an interval, whereas Cantor curves were defined in reference to the plane. The first satisfactory definition of a curve was established later in the 1920s in the framework of an even wider problem appertaining to dimension, and from that time onwards we understand a curve (not necessarily a plane curve) to mean a continuum of dimension 1 [C: Menger, 1932]. Even so, that did not diminish the significance of Sierpiński's curves.

Z. Janiszewski wrote his first and most important papers in Paris, including his excellent doctoral thesis [A: Janiszewski, 1911], containing results which are now classical: the theorem stating that if a continuum has points of intersection with a given set, as well as points of intersection with its complement, then it must have points of intersection with the boundary of the set; a topological characterization of an arc (an arc is an irreducible continuum that contains no proper subcontinuum of condensation). However, his most important result turned out to be a theorem, published in Polish with a French abstract [A: Janiszewski, 1913] and known today as Janiszewski's theorem: given two continua in the plane, neither of which divides the plane, their union does divide the plane if and only if their intersection is disconnected. The theorem identified a characteristic property of the plane. Its significance is that it considerably simplifies and shortens a proof of Jordan's theorem about dividing the plane with a simple closed curve (see [C: Kuratowski, 1952; pp. 358–359]). Furthermore, it turned out to be an important element in the topological characterization of the plane (see [C: Kuratowski, 1952; p. 374]). S. Mazurkiewicz published his first paper (on irreducible continua) shortly before Z. Janiszewski [A: Mazurkiewicz, 1910]. But his greatest success during those years was his characterization of images of the interval under continuous mappings, namely Peano continua (named in honor of G. Peano, the first person to discover that the square is such an image), by considering their local and global connectivity ([A: Mazurkiewicz, 1913], written in Polish, but with a French abstract). H. Hahn obtained analogous results in Czerniowce, as did R. L. Moore in the USA., but once Mazurkiewicz had presented his results in the French language in Volume I of "Fundamenta Mathematicae" [A: Mazurkiewicz, 1920], his contribution was acknowledged, and the characterization is known now as the Mazurkiewicz-Moore-Menger theorem (see [C: Kuratowski, 1952; p. 182]).

The earlier papers of S. Ruziewicz covered a not dissimilar subject matter concerning the singularities of real-valued functions. It was a popular topic at the time, motivated by an interest in acquiring the broadest possible understanding of what constituted real-valued functions. His first paper appeared in French but was issued by a publisher having little worldwide status [A: Ruziewicz, 1911]. Other subsequent papers were written in Polish, and so they did not become widely known. In his doctoral thesis [A: Ruziewicz, 1913], he constructed a continuous, monotonic function having no derivative at uncountably many points. It was a topic he would return to on more than one occasion, such as in [A: Ruziewicz, 1916], and it provided the basis for his later habilitation degree, obtained at Lvov University in 1918.

And so, even before World War I, a group of young, dynamic mathematicians had assembled in Lvov under the initiative and leadership of W. Sierpiński. The group worked on what was then the very new subject of set theory, but it had a special predilection for the theory of real-valued functions and for point set topology (now we would rather say general topology). The group sometimes published abroad. It marked a new moment for mathematics in Poland. Considering the enthusiasm of these people—their youth, their desire for results, as well as the sweet taste of first successes—one may see in it a highly auspicious beginning. However, there would be no subsequent follow up.

CHAPTER 4

The University in Warsaw and Janiszewski's Program (1915–1920)

With the outbreak of World War I Sierpiński's group in Lvov got broken up. Sierpiński himself, as an Austrian citizen, got interned in Russia, where he just happened to be spending his vacation. He was first sent to Viatka and later to Moscow, where he undertook joint research with N. Lusin[27]. Z. Janiszewski signed up as a volunteer with the Polish artillery corps and spent the first winter of the war on the Carpathian front. On June 5, 1918, he was formally invited to accept a chair at the newly reactivated University of Warsaw[28]. S. Mazurkiewicz also returned to Warsaw and took up the second mathematics chair (Janiszewski held the first) at Warsaw University. W. Sierpiński returned to his homeland on 1917. He was briefly in Lvov, but he certainly took up the third mathematics chair at the university in Warsaw and stayed there permanently. Only S. Ruziewicz stayed in Lvov.

Sierpiński's labors, and those of his group, were not, however, in vain, but we must briefly transport ourselves back in time to Warsaw—with Z. Janiszewski, S. Mazurkiewicz, and W. Sierpiński—because subsequent events there had an enormous influence on mathematics in Poland and on Lvov in particular.

The most important event was the reactivation of the university in Warsaw in 1915. K. Kuratowski was one of its very first students. Years later he recalled [B: Kuratowski, 1981; pp. 29–30]:

> It felt like our university was [...] a continuation of the Warsaw Main School that had been closed down half a century earlier by the Tsarist authorities. That link was made even more spectacular by the fact that among the professors were old heads of departments, former professors of the Main School, such as J. de

[27]Nikolai Lusin (1883–1950), a prominent representative of the Moscow school of the theory of functions, which enjoyed strong links with French mathematics. In 1936 he was the victim of Stalinist provocation in what became known as "the Lusin affair" (more precisely, the "affair of the academic Lusin"): it was a consequence of the heavy-handed censorship of Soviet scholarship, including mathematics, enforced by Soviet authorities. Lusin managed to save his own life but there were many other victims, found guilty of "Lusinshchina", who did not. During his time in Moscow, Sierpiński wrote four joint works with Lusin and later organized an international appeal from the mathematical community in his defence. See [B: Duda, 2001], [B: Duda, 2004a]. Bolesław Młodziejewski (1858–1923), a mathematician of Polish extraction and a professor at Moscow University, played an influential role in getting Sierpiński moved from Viatka to Moscow.

[28]In the summer of 1915 the Imperial University in Warsaw was evacuated to Rostov-on-Don. A few months later a Polish university was set up in Warsaw, upholding the same old traditions which had existed there between 1862–1869. Janiszewski was actively involved with the Polish university from its very inception.

> Baudoin de Courtenay, W. Miklaszewski, not to mention the former students of that school. [...] And so the opening of the university [...] was of much greater national significance than just the academic appointments themselves. Warsaw too understood that the university served this role. The university became the subject of wide interest, bordering on cult interest. In lectures one could meet [...] many broad-minded citizens having many interests, who wanted to immerse themselves behind university walls, which was something previous generations had fought so hard for. [...] The list of professors was of a particularly high standard. Besides the professors of the Free Warsaw Common School (a substitute university in the times before World War I), and in addition to the professors from Cracow and Lvov, many scholars who had previously held other university chairs abroad were appointed to chairs at Warsaw. Among the latter group were the physicist [Józef] Wierusz-Kowalski from Switzerland, the lawyer [Leon] Petrażycki and the classicist Tadeusz Zieliński—both from Russia, the expert on Roman law [Ignacy] Koschembar-Łyskowski from Germany, a little later there arrived Stefan Pieńkowski, a physicist working in Germany, as well as many others.

Z. Janiszewski, one of the first professors of the university in Warsaw, wrote an article [B: Janiszewski, 1917] in response to an appeal made in a special volume, prepared by the Mianowski Fund, regarding the needs for science in a reemergent Poland. Despite its concision, that article laid out a program that was followed by most of the next generation of Polish mathematicians. And not just Polish mathematicians. For example, here is how J.-P. Kahane concludes his article on the Polish School of Mathematics during the years 1918–1939: "We all owe a great debt of gratitude to that great mathematician, democrat, and patriot. [...] I shall finish [with this] advice to mathematicians and nonmathematicians alike: read Janiszewski" [B: Kahane, 1995; p. 175].

Even today one is struck by the depth and originality of Z. Janiszewski's strategic vision. Starting with an assessment of the situation at the time, he perceived there was a possible way for "Polish mathematics to achieve independent status". The plan was to: select a single area of mathematics (the natural candidate being set theory and allied fields where set-theoretic methods played an important role, such as the theory of functions and topology—the area of interest which had been pursued by the now-disbanded Lvov group led by Sierpiński) for the majority of Polish mathematicians to focus on; foster a culture of mutual research collaboration, providing support for younger members; establish a journal dedicated exclusively to the selected area of interest and in which all articles had to be published in a language used at mathematical congresses.

The proposed program was undoubtedly shocking. If Polish mathematicians were to focus most of their efforts exclusively on a single mathematical field, there was a very real danger other areas would be neglected, including fundamentally important, classical mathematical fields, such as analysis, geometry, and algebra. Overcoming these very real dangers, and successfully finding a way for "Polish mathematics to achieve independent status", required a cold assessment of some harsh realities: the absence of any serious home tradition in mathematics; extremely poor library resources, which were totally inadequate for supporting long-term studies; and one also had to consider the woefully small number of faculty members among the younger generation of mathematicians. The decision to confine the program to a single academic field shows us not only how realistic Janiszewski's assessments were. It equally reflects the experience and maturity of judgement representative of Sierpinski's Lvov group, most particularly its culture of mutual cooperation and exchange of ideas, conducive to the significant originality of works produced by its members.

A totally original feature of Janiszewski's plan was the idea of establishing a journal to serve the needs of an as-then nonexistent group (when the first volume of Fundamenta Mathematicae appears a few years later, it will contain twenty-five articles from barely eight Polish mathematicians, of whom just four came from Warsaw). A journal dedicated to a single mathematical field, and a new one at that, seemed doomed to failure from the outset because never before had there been a journal of such narrow scope. These were powerful arguments against the program, and criticism of it emerged both at home and abroad[29]. One may further add the wounding of national pride at the prospect of abandoning use of the Polish language.

However, conditions at the time were favorable. Support for the program came from the reactivated Warsaw University, where a young generation of scholars (Z. Janiszewski and S. Mazurkiewicz) and their students (B. Knaster, K. Kuratowski, and others) were highly enthusiastic about it, brimming full of confidence in their own abilities and for the future. The vision was shared by W. Sierpiński, who had returned from Russia and who, in 1918, had assumed the third chair of mathematics at Warsaw University:

> When the three of us, Janiszewski, Mazurkiewicz, and I, found ourselves as professors in Warsaw at the reborn university, we decided we would realize Janiszewski's plan, to publish a journal in Warsaw, appearing in various languages, dedicated to set theory, topology, the theory of real functions, and mathematical logic. That is how "Fundamenta Mathematicae" came into being. [B: Sierpiński, 1963], [B: Sierpiński, 1967]

Thus began the Warsaw School of Mathematics. It focused on "set theory and its applications" (as stated in the header on the cover of "Fundamenta Mathematicae"), or more precisely, on the pure theory of sets, set-theoretic topology, the theory of functions of a real variable, and mathematical logic[30]. In mathematics, the leaders were W. Sierpiński and S. Mazurkiewicz (later joined by younger people: K. Kuratowski, S. Saks, and others). In logic, the leaders were S. Leśniewski and J. Łukasiewicz.

The close working links with logic were yet another characteristic feature of the Warsaw school . They had their origin in Lvov, where K. Twardowski's superb school[31] of philosophy was run at the turn of the twentieth century; it placed great emphasis on precise statements and clarity of thought. These links meant much

---

[29]In Poland, there was opposition from Stanisław Zaremba in Cracow, while from abroad there was criticism from Henri Lebesgue and Nikolai Łusin; see [B: Duda, 1995a], [B: Duda, 1996].

[30]See [B: Duda, 1996].

[31]See [B: Woleński, 1985], [B: Woleński, 1988].

to both J. Puzyna and W. Sierpiński, and hence to their students too. Not surprisingly, when, after 1915, some of K. Twardowski's and some of W. Sierpiński's students found themselves in Warsaw, regular meetings took place between mathematicians and logicians, including Z. Janiszewski, T. Kotarbiński, S. Leśniewski, J. Łukasiewicz, S. Mazurkiewicz, W. Sierpiński (after 1918), and others. Friendships were formed and mutual cooperation ensued, with outstanding results, so outstanding that logic advanced from being a peripheral discipline of philosophy to a major branch not only of philosophy but of mathematics too. The Polish School of Mathematics played a decisive role in this advancement during the interwar years and made a significant contribution to mathematical logic around the world [B: Duda, 2004c].

CHAPTER 5

World Mathematics (Active Fields in Poland) Around 1920

Before we move on to further considerations, let us briefly acquaint ourselves with the state of mathematics around 1920, focusing on those areas where Polish mathematics would later distinguish itself so highly during the interwar years, in both its Varsovian and Lvovian guises. The relevant areas were the theory of functions, set theory, and topology (we shall describe them in that order), as well as probability theory (see Chapter 15) and mathematical logic.

One of the most important ideas of the new mathematics was the notion of a function. It entered into mathematics during the eighteenth century, but it only assumed greater generality and precision during the nineteenth century[32] at the end of which it had become a dominant notion not only in mathematical analysis but also in algebra (laws of composition) and geometry (transformations) and even in mathematical didactics[33].

Research into fundamental properties of functions, such as continuity, differentiability and series representation, etc., became known as the theory of functions. Sometimes reference was made to narrower theories, such as the theory of functions of a real variable, or the theory of analytic functions, etc. But when considered in its entirety, the theory of functions did not have clearly defined boundaries and it wandered into other territories, which in turn entailed new problems and new methods. Its characterizing feature was its strong influence on new ideas, such as Cantor's theory of sets or the topology of point sets. Consequently, the theory of functions became a lively area of interest, bountiful in unexpected theorems and difficult problems.

The notion of a function assumed an ever greater role and ever greater significance. Not only were individual functions investigated but so too were their sequences and, more generally, sets of functions[34]. Gradually the thought dawned, at first in a vague and intuitive sense only, that functions could be treated just as if they were elements ("points") belonging to these newly defined sets and that these sets might have curious properties, properties determined by the very functions which formed the sets themselves. Such sets were soon called "spaces". Rather more precisely, and in order to emphasize their special character, they were called "function spaces". The birth of these ideas seems vague and for a long time their actual terminology was treated merely as a convenient convention, but with the

[32]See [B: Youshkevich, 1976], [B: Felgner, 2002].

[33]An important part of the so-called Merans program, prepared in 1905 by the Commission for Mathematics Teaching under the guidance of F. Klein, was the emphasis on teaching in schools the general notion of a function (or a transformation, in the case of geometry).

[34]For further details and references about the evolution of research into sets of functions see [B: Duda, 2004b].

fullness of time the deeper advantages stemming from such practice became evident.

Notable contributions to these ideas were made in the nineteenth century by German mathematicians (above all by B. Riemann), the Italians (G. Ascoli, C. Arzelà, V. Volterra, and others), and the French (J. Hadamard, E. Borel, M. Fréchet, and others), together with the Swede E. I. Fredholm. Towards the end of this period the notion of a functional appeared for the first time, as a function assuming real values and defined on a given function space. Most particularly, there appeared the notion of a linear functional U satisfying the following condition:

$$U[\lambda_1 y_1(x) + \lambda_2 y_2(x)] = \lambda_1 U[y_1(x)] + \lambda_2 U[y_2(x)],$$

where λ_1, λ_2 are arbitrary real numbers and y_1, y_2 are arbitrary elements (functions) of the considered function space. Here, x is an arbitrary element of the set on which the functions in the space are defined. The limit of a sequence, the sum of a series, the limit of a function, the derivative, the integral, etc., are all examples of a linear functional (defined on an appropriate function space). This shows just how broad in scope this notion was and why there was good reason to believe it would become of major importance to mathematics.

The new concepts impacted particularly strongly on the development of integral equations. D. Hilbert picked up on the work undertaken on this by V. Volterra and E. I. Fredholm, publishing six papers on this topic during 1904–1910 [C: Hilbert, 1904–1910], and subsequently putting them together in book form [C: Hilbert, 1912]. Hilbert's motivating idea was to find solutions to integral equations, first by considering the solution sets of certain linear systems in many variables and then looking at the limit when the number of variables becomes arbitrarily large. However, his greatest success in this field is a general spectral theory for symmetric kernels (see [B: Steen, 1976], [B: Janas, 1993]). It pointed to the advantages flowing from the general approach and was a strong impetus for conducting further research on function spaces. It led to parallel progress in set theory and its associated topology. These three fields were hard to distinguish from each other at the time. And new terminology entered into the language of function spaces. Namely, functions that belonged to function spaces began to be referred to as vectors, while the original understanding about what constituted function spaces became refined through the discovery (or through the definition) of geometric structures lying in these spaces. The intention was to construct a framework for performing analysis on these spaces, a framework that required definitions for notions such as convergence, limit, continuity, etc. Hope was expressed that analysis of such a general nature would be capable of incorporating various branches of classical analysis, such as differential equations, variational calculus, integral equations, systems of linear equations in infinitely many variables, and so forth.

D. Hilbert's work was continued primarily by E. Schmidt and F. Riesz. The former considered D. Hilbert's so-called *sequence space*, that is, a space consisting of sequences of complex numbers $(z_1, z_2, \ldots)$ and such that $\sum_{i=1}^{\infty} |z_i|^2 < \infty$. He treated it as an independent object in its own right and endowed the space with notions and methods borrowed from Euclidean geometry. That is how there arose notions (the names were given later) such as the *scalar product*[35] of two vectors

[35]The *scalar product* of two vectors $(z_1, z_2, \ldots)$ and $(w_1, w_2, \ldots)$ is defined as the number $(z|w) = \sum z_i \bar{w}_i$.

lying in the space (i.e. a space of sequences), *orthogonality* of two vectors[36], *vector norms*[37], *strong Cauchy sequences*[38], *strong algebraic closure*, etc., as did other results such as the Schwarz inequality and the projection theorem. Hilbert's sequence space, denoted ℓ^2, turned out to be a richly rewarding and highly interesting one.

Further function spaces followed on after the space ℓ^2, together with highly unexpected correspondences. The first correspondence, a most important one, was discovered by E. Schmidt and M. Fréchet. They showed (independently of each other) that the set L^2, comprising square-summable functions (in the sense of Lebesgue integration), has the same geometry as Hilbert's sequence space ℓ^2 [C: Fréchet, 1907]. This astonishing result underlined the existence of deep links between different function spaces because it showed how two function spaces can be defined in completely different ways but can still be identical to each other in a certain sense. The identical structure of these two spaces was a powerful impetus for seeking a general structure, or rather to define a general "function space" which could include the two aforementioned spaces, and others, as special cases.

F. Riesz similarly delivered many interesting notions and results in the decade immediately preceding World War I. Among other things, he introduced a whole class of new spaces L^p, $1 < p < \infty$, endowing each of them with a metric[39]

$$d(f,g) = \left\{ \int_a^b |f(x) - g(x)|^p dx \right\}^{1/p},$$

with respect to which all the spaces are complete [C: Riesz, 1910]. He proved the wonderful representation theorem and introduced the notions of dual operator, transpose operator, etc. He even tried (unsuccessfully) to develop a general approach to function spaces. Up until 1920, when the decisive breakthrough came, many function spaces had been identified and much research had been done on them. Nonetheless, it was a disordered treasure-trove lacking clear pathways through it and lacking a clear, general overview.

Early attempts to establish an analysis theory for function spaces were summarized in the book [C: Lévy, 1922], from which it began to be called functional analysis. A summary of the analysis associated with the theory of linear equations and quadratic forms in infinitely many variables was given in the article [C: Hellinger & Toeplitz, 1928].

The appearance of problems in the theory of function spaces and their subsequent development were greatly influenced by topology and by the theory of point sets, a subject which had arisen slightly earlier. The theory of sets was attributed to a single individual, G. Cantor. When this graduate of the Berlin school of Weierstrassian analysis arrived in Halle (where he remained for the rest of his life), he investigated, under H. Heine's guidance, the uniqueness question for trigonometric sums: if a series vanishes at every point lying outside a certain set Z, is it true that all its coefficients are zero? Cantor started by showing the answer is affirmative if the set Z consists of finitely many points. But he didn't stop there, because

[36]Two vectors in a sequence space are said to be *orthogonal* if their scalar product is zero.

[37]The *norm* of a vector z in a sequence space is defined as the number $|z| = (\sum z_i \bar{z}_i)^{1/2}$.

[38]A sequence of vectors $z_1, z_2, \dots$ in a sequence space is said to be a *strong Cauchy sequence* if $\lim |z_i - z_j| = 0$. Every strong Cauchy sequence is *strongly convergent* to some point z in the given space; i.e. $\lim |z_i - z| = 0$.

[39]The notion of a metric was first introduced (but called something else) in the work [C: Fréchet, 1906].

he already had examples showing the answer to be affirmative for certain sets Z having infinitely many points. By exploiting such generalizations, he laboriously went about proving uniqueness theorems and setting up requisite mathematical tools for describing certain sets on the real line, beginning with the notion of a limit point of a set Z, the set Z' of all limit points of the set Z, in other words the so-called derivative of the set Z, closed sets, etc. [C: Cantor, 1932]. His subsequent papers generalized these concepts, extending them from the line to n-dimensional Euclidean space. A new world then opened up for mathematicians. Some embraced it enthusiastically while others were repulsed: either way its influence on mathematics was immense. Although various threads wove their way through this new mathematical world, it was still referred to as the theory of sets[40]. It was not until the 1930s when it began to be divided up into particular parts: the pure theory of sets, with a tendency to treat it as the foundation of all mathematics; the topology of point sets (nowadays general topology), which became the language of most mathematics; the very lively theory of functions; and others.

Although the first clearly defined notions of general topology arose from Cantor's work, its true origins are actually older. Already as early as the seventeenth century, G. W. Leibniz proposed the development of *Analysis situs* as a branch of mathematics in which measurable quantities (e.g. lengths, angles, areas) were ignored, the important considerations being quantities that were independent of position or shape. The idea was rather vague but it inspired many mathematicians. That idea assumed clearer form when G. Cantor began his research into subsets of Euclidean space, leading to the topology of sets (general topology), and to H. Poincaré's thesis [C: Poincare, 1895], from which algebraic topology was born. Almost from the very beginning topology split into two paths: algebraic topology and set topology. In the ensuing years, the latter topic would have incomparably greater significance for the Polish mathematical school (including the one in Lvov) than the former. A state-of-the-art summary of the subject was given in the monograph [C: Hausdorff, 1914] (see [C: Hausdorff, 2002] for the text with commentary), but because of the war its effective influence dates from 1920 onwards.

Polish mathematicians alerted others to their presence in each of the three aforementioned mathematical fields, fields which in the 1920s were still not considered to be distinct from each other. Most prominent of all those working on the theory of sets, topology, and the theory of functions were W. Sierpiński and his group: Z. Janiszewski, S. Mazurkiewicz, and S. Ruziewicz (see Chapter 3). H. Steinhaus made the most notable contribution to functional analysis (see Chapter 6).

---

[40] *Mengenlehre, Theory of Sets, Théorie des Ensembles.* In his treatise [A: Puzyna, 1898–1900], the author used the expression *the theory of sets* ["teoria mnogości"], which was commonly adopted in Poland.

Part II. THE GOLDEN AGE
Individuals and Community

CHAPTER 6

The Mathematical Community in Lvov after World War I

Mathematics in Lvov sprang into life at the same time that the Warsaw School of Mathematics was starting up. It did so over the rubble of three wars (World War I, the Polish-Ukrainian War, and the Russo-Polish War), which had been ruinous for mathematics in Lvov. Sierpiński had left Lvov. Let us recall that he had been interned by the Russians during 1914–1917. It is true that he came to Lvov in 1917 for a few months and that he lectured there during the summer semester 1917/1918, but he had *de facto* moved to Warsaw, where he assumed the third mathematics chair. Z. Janiszewski and S. Mazurkiewicz had already left earlier, and the two of them asssumed the first two mathematics chairs at Warsaw University (though the former died at the beginning of 1920). S. Ruziewicz was the only one of Sierpiński's original group still remaining in Lvov. Among the mathematicians still active there were A. Łomnicki and J. Puzyna.

However, changes were happening. The Lvov Mathematical Society, with J. Puzyna as its appointed president, was established as early as 1917, despite the difficult war conditions and thanks to the efforts of Z. Janiszewski. While visiting Cracow in that same year of 1917, J. Puzyna persuaded H. Steinhaus, at the time a "private scholar", to go and deliver his habilitation lecture at Lvov University. A year later S. Ruziewicz would also obtain his habilitation degree. The Lvov Mathematical Society was not making much progress, and then J. Puzyna died in 1919. After W. Sierpiński's departure and J. Puzyna's death, the two mathematics chairs at Lvov University remained vacant. Military and political events saw Lvov become part of Poland, but the city was Polish not only because of its formal territorial status. It had a dominant Polish population and a predominantly Polish culture (in 1938, 63.5% of Lvov's inhabitants spoke Polish as their native language, while 85% spoke it on a daily basis).

Despite its predominantly Polish character, the city was nonetheless multilingual and multicultural. Ukrainians considered it to be just as much their city too. There were not that many of them (in 1938 only 11.3% of inhabitants declared Ukrainian as their native language), but they did constitute an important part of the population, with aspirations of their own. The Polish authorities tried to curb Ukrainian ambitions. For example, they closed down the unofficial Ukrainian University and did not allow it to continue in any official capacity. However, the Ukrainian Science Society, named after Taras Szewczenko, did function, albeit as a substitute for the nonexistent Ukrainian Academy of Sciences. There were Ukrainian secondary schools and high schools. Books and journals in the Ukrainian language were also produced. The first Ukrainian mathematicians were there: M. Czajkowski (who later went to Soviet Ukraine and got sent to a labor camp), W. Lewicki (the author of the first mathematical work in Ukrainian), M. Zaryckij (H. Steinhaus thought highly of him and persuaded him to move to Lvov in 1925). They taught in high schools, with a preference for Ukrainian ones, they lectured at the underground Ukrainian University until it was closed down in 1924, and they were active in the Taras Szewczenko Science Society. M. Zaryckij was the only one of those three who could genuinely claim to belong to the Lvov mathematical school. While life in Lvov and Eastern Galicia was better for Ukrainians from an aspirational perspective in comparison with Soviet Ukraine, where Ukrainian "nationalist" sentiments were becoming ever more ruthlessly stamped down, Polish-Ukrainian tensions did exist and became dramatically evident during World War II.

All that was to come later (see Chapter 25 in particular), but back in 1920 peace had at last come and Lvov University, having been changed to the Jan Kazimierz Univeristy of Lvov (abbreviated to UJK), started afresh, as did its mathematical section. Its leading figure was to be H. Steinhaus. Hugo Steinhaus came from Jasło. After passing his matura exam with distinction in 1905, at the high school there, he started studies in mathematics and philosophy at Lvov University, where he attended lectures given by J. Puzyna and K. Twardowski. But after his first year of study, and at the instigation of a family friend, Professor S. Jolles from Charlottenburg Polytechnic (Berlin), he moved to Göttingen, at that time the Mecca of mathematics. D. Hilbert, F. Klein, H. Minkowski, H. Weyl, E. Landau, among others, gave lectures there. In 1911 he passed his rigorosum before a commission consisting of D. Hilbert, C. Runge, and P. Hartmann and was awarded the title of Doctor of Philosophy *summa cum laude* for his thesis on Dirichlet's principle [A: Steinhaus, 1911]. Upon obtaining his doctorate he returned to his homeland and worked as a "private scholar". He published some notes in Polish journals, including a construction, remarkable for its time, concerning a trigonometric series whose terms converge to zero but which diverges everywhere [A: Steinhaus, 1912].

During 1914–1918 Steinhaus served as a soldier in World War I (he took part in the Wolhynia Campaign of 1915 as a volunteer legionary), was a clerk in the Central Office for Country Reconstruction, and also pursued academic work. Encouraged by J. Puzyna, he did his habilitation in February 1917 at the university in Lvov. The procedure looked much as it does today: first the departmental council heard J. Puzyna's talk detailing his academic achievements, there then followed a habilitation colloquium, and at the end H. Steinhaus delivered a so-called inaugural lecture. His was entitled "Zwodnicze Drogi Matematyki" [Misleading Mathematical Pathways]. After that he was awarded his habilitation degree. With it, he obtained his *veniam legendi*, authorizing him to lecture. He consequently left the Central Office in Cracow and moved to its branch in Lvov. He considered himself a poor clerk because he was much more interested in lectures and doing mathematical research, but from May 1, 1918, he had a permanent assistantship position. When the Polish-Ukrainian War of 1918 broke out, he moved back to more peaceful Jasło, and from there he returned in 1920 to accept a position as professor extraordinary at UJK (he became an ordinary professor there in 1923). The quarter-century 1917–1941 he spent at Lvov was the happiest and most productive years of his life. Only with the outbreak of the German-Soviet War in 1941 was he ousted from Lvov.

FIGURE 2. This photo, taken in 1898, shows the Diet of Galicia province, then part of the Austro-Hungarian empire: it became the main building of Lvov University after World War I.

H. Steinhaus's predecessor as mathematics professor at Lvov University was Eustachy Żyliński, who had come to Lvov in 1919 after J. Puzyna's death. They did not really understand one another; Żyliński was a professor from an older era. He focused on teaching, whereas Steinhaus, having attended Göttingen University, believed mathematicians are only as good as the theorems they prove. Nonetheless, they did tolerate each other. Steinhaus became more active both researchwise and in his teaching responsibilities. He was the first Polish mathematician to write a paper on the newly emerging field of functional analysis, finding a canonical representation of a functional in the space of Lebesgue integrable functions [A: Steinhaus, 1919][41]. Another paper that he wrote around that time [A: Steinhaus, 1920], concerning the set of distances between points lying in a set of positive measure, was often cited later.

H. Steinhaus was a great scholar, but he considered his greatest mathematical discovery to be ... Stefan Banach. It began thus:

> One summer, in 1916, as I was walking along the "Planty"[42], I overheard a conversation, or rather only a few words. I was so struck by the words "the Lebesgue integral" that I came nearer to the bench on which the speakers were sitting and, then and there, I made their acquaintance. The speakers, Stefan Banach and Otton Nikodym, were discussing mathematics. They told

[41]This work entered the history books of functional analysis; see [B: Dieudonné, 1981; p. 128].

[42]The Planty is a garden park area surrounding the old city of Cracow.

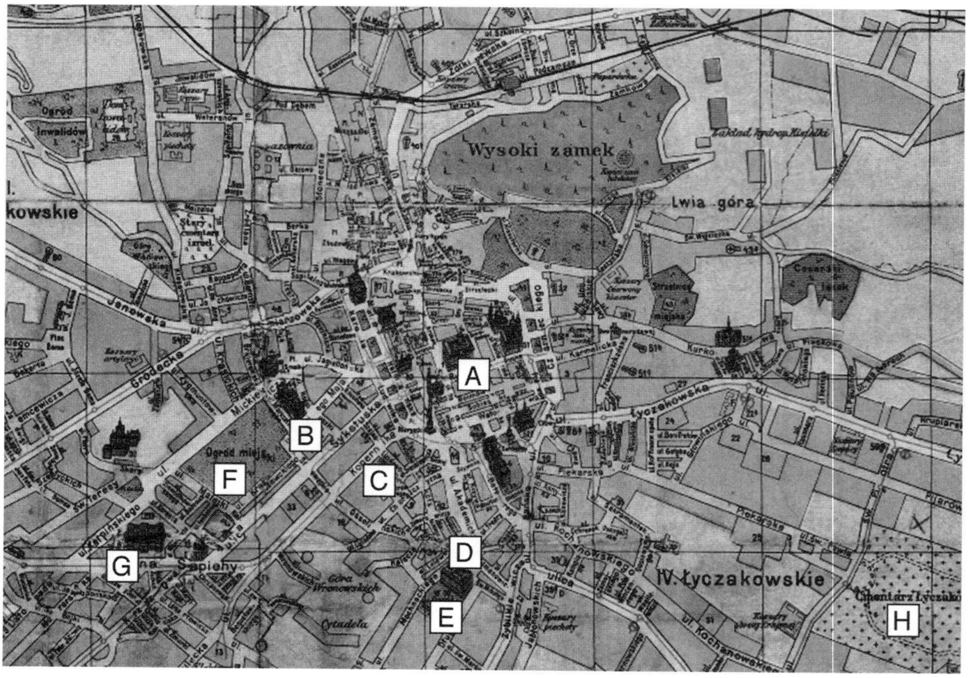

FIGURE 3. Grand plan of the center of Lvov, 1910: A—the old city, B—the main building of UJK, C—the Ossolineum, D—Fredro Square (where the Roma Café and the Scottish Café were located), E—the "Old 'Versity", F—Kościuszko Park, G—the main building of Lvov Polytechnic (PL), H—the Łyczakowski Cemetery (where Banach's grave is located).

me they had another chum—[Witold] Wilkosz. It was not only mathematics that bound together those three young men, it was the hopeless situation of young people in the "fortress Cracow" (such was the official status of Cracow in those days of war), the insecurity of the future, the difficulty of earning one's living, the lack of contacts not only with foreign scientists, but even with the Polish ones—such was the atmosphere of this city in 1916. But all that did not prevent the three young men from spending a lot of time in cafés discussing mathematical problems amidst a noisy crowd.

[B: Steinhaus, 2010; p. 93]

From that time we met regularly. And since in Cracow there were also Władysław Ślebodziński, Leon Chwistek, Jan Króo, and Włodzimierz Stożek, we decided to establish a mathematical society [...]. The society was developed.

[B: Steinhaus, 2010; p. 93]

Settling down in Lvov, H. Steinhaus arranged for S. Banach to become an assistant under A. Łomnicki at the polytechnic and later helped him obtain his doctorate (which was not easy as Banach had only studied for two years at Lvov Polytechnic,

FIGURE 4. "Old 'Versity" beside St. Nicholas Church.

meaning he had not completed a formal course of higher study), his habilitation at UJK, and a position as professor extraordinary in 1923. S. Ruziewicz likewise progressed professionally: as of January 1, 1921, he was made professor extraordinary and a full professor at UJK in 1924. In 1924 the UJK philosophy department was split into the Department of Humanities and the Department of Mathematics and Natural Sciences. In 1926 UJK introduced a master's course. At the Department of Mathematics and Natural Sciences there were four mathematics chairs, held by the following four professors (named in the order of the appointments): Eustachy Żyliński, Hugo Steinhaus, Stanisław Ruziewicz, and Stefan Banach.

A contemporary student, K. Szałajko, describes the mathematics staff at UJK and the topography of the place at the start of the 1930s:

> The main building of Jan Kazimierz University [Figure 2] was located on Marszałkowska Street, at the edge of Kościuszko Park. The university authorities had their offices there, meaning the rector, the deans, clerks, part of the law department, the chairs of philosophy, the theology department and part of the humanities department. It also included lecture rooms, including one of the largest, the Copernicus lecture hall, where some mathematics and physics lectures were given, and where, in particular, Professor Kazimierz Ajdukiewicz, director of one of the two humanities departments, delivered his lectures in philosophy and logic. The annual inaugurations took place in the beautiful lecture theater in the main building.
>
> Going along Słowacki Street, Ossoliński Street, and a few other minor ones led to Fredro Square, where there were two cafés situated close to each other: one was called "Roma"; the

other would become known in the mathematical world as the "Scottish Café" [Figures 10 and 11]. The opposite side of Fredro Square led on to St. Nicholas Street, where standing next to the Church of St. Nicholas, on a small hill, was the old university building known as the "Old 'Versity" [Figure 4]. Here, leading on from the stairs on the first floor along a corridor and passing two "mathematical" lecture rooms, one came across another corridor, perpendicular to the first, which led directly to a room used by professors and mathematics students. Professor Eustachy Żyliński, the head of Mathematics Chair A, had his office in a corner room, next to which was the room for assistants and the mathematics library, which housed a large collection of books. Next in the enfilade was a student reading room, then the journals reading room and the office of Professor Stanisław Ruziewicz, head of Mathematics Chair D. The two offices further on belonged to Professor Stefan Banach, head of Mathematics Chair C, and Professor Hugo Steinhaus, head of Mathematics Chair B. The offices of Professors Żyliński and Ruziewicz could be entered directly from the corridor, whereas those of Professors Banach and Steinhaus were accessible from the corridor via a small anteroom. On the second floor there were two rooms for readers. And that was all that was available to the four mathematics chairs of Lvov University.

The building referred to above had other lecture rooms and was where other chairs of the Department of Mathematics and Natural Sciences were situated. Two rooms on the first floor were reserved exclusively for mathematics lectures and equally for students of mathematics, physics, and astronomy. Lecture Room 14, the largest in the "Old University", was on the second floor and was where, among other things, Professor Kazimierz Twardowski gave his philosophy lectures, which were immensely popular with students and attendees "from the town". Professor Leon Chwistek, who came to Lvov from Cracow, also lectured there. His lectures were also very well received.

Very near to the old university building, namely on Długosz Street, a side street of St. Nicholas Street, were the physics and chemistry buildings. On Maurice Mochnacki Street, at the other side of the old building, was the university library, on which, engraved at the front, were the words ARS LONGA, VITA BREVIS [Art is long, life is short]. A botanical garden stood between Mochnacki Street, Mikołaj Street, and Długosz Street. Part of it was accessible as a park, where students from the Department of Mathematics and Natural Sciences would go for a walk. [...]

A mathematics library and a student reading room, serving the needs of a broad program of mathematical studies, were commonly available to everyone from any of the four mathematics

FIGURE 5. The main building of Lvov Polytechnic, the view of the façade seen from Leon Sapieha street. The Church of St. Mary Magdalene lies in the background.

chairs. Together they formed the so-called mathematical seminar. One could effectively stay all day in the mathematical seminar, consulting books lent by assistants or student assistants on duty between the hours 8:00 am–1:00 pm and 3:00 pm–8:00 pm. Taking books home was forbidden. Furthermore, it was rare for there to be two copies or more of the same textbook, and even then the number was very limited, so students had to come early if they needed a particular book. The reading room was often full because for most people it was the only opportunity to study textbooks. Going through the student reading room led to the room for assistants, while going in the opposite direction led to the journals reading room. Here was the journals library, where journals were laid out on open shelves. The journals reading room was mainly used by academics and older students preparing for their diplomas. The journals reading room was also used for seminars and meetings of the Polish Mathematical Society.

[B: Szałajko, 1984; pp. 85–87]

Readers had one or two rooms on floor II. The "mathematical seminar" was assigned only one "mathematical" lecture room, having the number 2, while others could be "borrowed" when needed. As one can see from this account, the space available for mathematics was modest, but so too was the number of salaried positions. There was only one salaried position as an assistant professor for all four mathematics chairs, serving Mathematics

> Chair A and it was held by Reader Herman Auerbach, as well as some other positions for younger and older assistants.
>
> [B: Szałajko, 1994; p. 252]

The main UJK building looked onto Kościuszko Park, and, situated on the other side, on L. Sapieha Street, was Lvov Polytechnic, denoted PL [Figure 5]. A singular feature of PL was its placement, next to four technical departments, the agro-forestry department, and the general studies department. Mathematical chairs at PL were allocated to departments. Mathematics Chair I was allocated to the Department of Land and Water Engineering, Mathematics Chair II to the mechanics department, Mathematics Chair III to the general studies department. Responsible for these chairs were Placyd Dziwiński (until his retirement in 1925: Chair I), Włodzimierz Stożek (between 1922–1941; first with Chair III, later with Chair I), Antoni Łomnicki (between 1919–1941; Chair II) and Kazimierz Kuratowski (between 1927–1933, i.e. until the liquidation of the general studies department; Chair III). Besides this there was the Department of Descriptive Geometry, run by Kazimierz Bartel. From 1937 Kazimierz Szałajko was a senior assistant at Antoni Łomnicki's chair, while Andrzej Turowicz was senior assistant at Włodzimierz Stożek's chair, and Stanisław Mazur was an assistant professor or adjunct.

> [Professor Łomnicki] was above all a teacher and expositor of applied mathematics. His three-volume textbook entitled *Differential and Integral Calculus*, having the subtitle "for natural scientists and technicians", was, it is fair to say, a pioneering work [...]. Profesor Łomnicki was the first to initiate a mathematics lab into the mathematics teaching and higher technical schools, having of course only modest means at his disposal, and he engaged a whole team of mathematicians, taken from both chairs, to run it. Thanks to the professor's efforts there were many pictures stuck to boards, cartographic projections, and nomograms showing nets and ladders. He used those boards for his lectures. As a mathematical cartographer he was known across Europe, and therefore a member of the Paris-based International Cartography Society. Łomnicki was one of the first [...] pioneers of axiomatic probability theory. And who could have been unfamiliar with his geometry textbooks?
>
> [B: Szałajko, 1994; p. 262]

The task of the PL general studies department was to prepare students in the other departments with basic knowledge (mathematics, physics, etc.) and to train teachers of technical subjects. Its benefit for mathematics was that it offered more jobs to mathematicians. However, Minister Jędrzejewicz had it abolished on September 25, 1933, and Kazimierz Kuratowski returned to Warsaw, where he was made an ordinary professor at the university there.

K. Kuratowski had a very productive time in Lvov, during which he published some forty papers, including joint papers with S. Banach and S. Ulam. His tutees included S. Ulam and E. Otto.

The PL general studies department worked closely with UJK mathematicians. Apart from professors from the politechnic (K. Bartel, K. Kuratowski, A. Łomnicki, W. Stożek), professors and readers from the university (S. Banach, S. Kaczmarz,

W. Nikliborc, W. Orlicz, S. Ruziewicz) also lectured there. It was a superb faculty. S. Banach's interest in mechanics doubtless dates to the time he spent meeting the needs of that department, as shown by the monograph he wrote [A: Banach, 1938b].

There were few positions with paid employment at UJK and PL, obtaining one was difficult, and the pay was low. Herman Auerbach is an illustrative case. Upon completing his mathematics studies at UJK in 1923 he obtained a position as a "demonstrator for the mathematical seminar" for the year 1923/1924 for which he was required to work twenty hours per week on a salary 50% lower than an assistant in [salary] group VIII, amounting to 130 zloties per month. The pay and conditions changed almost every year (as did "daily employment"), but his salary was always low and his hours long. The responsibilities of a demonstrator were not precisely stated, so one can safely assume he ended up doing just about everything asked of him, whether it was administrative work or lecturing on articles for seminars as requested by the professors. Employment on the lower rungs of the academic ladder meant a yearly contract. In H. Auerbach's case, the department applied to the UJK Academic Senate every May or June to have it renewed for the subsequent year, which it always was, allowing him to slowly progress professionally. So for the two years 1923/1925 he was a "demonstrator" and a "junior assistant" on a group IXb salary for the years 1925–1930. In 1930, when he got his doctorate, he was promoted to "senior assistant at Mathematics Chair A", on a group VIII salary. And thus it continued. Even when he obtained his habilitation in 1935, thereby authorizing him to deliver lectures, he was still registered as a senior assistant, the difference being that as a reader he received a group VII salary, i.e. 335 zloties per month. It still wasn't much, because he admitted he was earning about 130 zloties per month extra at a private junior high school. The last document in his personal file is dated July 24, 1939, and is a nomination, signed by the UJK rector, for H. Auerbach to be appointed as senior assistant for a one-year period September 1, 1939–September 30, 1940. The dean submitted the nomination "for the attention of the director of Mathematics Section A" on September 1, 1939, but that was the very day war broke out: Germany had invaded Poland.

The list for the mathematics chairs at UJK for the academic years 1936/1937 and 1937/1938 reads as follows:

Chair A. Head: Professor Dr. Eustachy Żyliński. Senior assistant: Reader. Dr. Herman Auerbach.

Chair B. Head: Professor Dr. Hugo Steinhaus. Senior assistant: Reader. Dr. Juliusz Schauder. Deputy assistant (volunteer): Helena Plamitzer.

Chair C. Head: Professor Dr. Stefan Banach. Deputy assistant: Kazimierz Szałajko MSc. Senior messenger: Józef Góral.

CHAPTER 7

Mathematical Studies and Students

The master's-level courses (magisterium) introduced to Polish universities in 1926 were designated to last four years. Each academic year was broken up into trimesters (later into semesters), with the fourth year reserved for seminars, writing theses, and final examinations.

Preserved in the Lvov archives are the files documenting the progress of the student Stanisław Ulam, who spend his first year of studies, in the "scientific year" 1927/1928, at the PL mechanics department. He spent the next three years in the mathematics group at the PL general studies department. He wrote down his schedule for the year 1929/1930:

Mathematics III (Professor Stożek, 1 hour in the winter semester/0 hours in the summer semester).

Problem classes in mathematics III (Professor Stożek, 2/0).

Topology (Professor Kuratowski, 2/2).

Mathematics seminar (Professor Kuratowski, 1/1).

Analytic functions (Professor Ruziewicz, 3/3).

Applications of integral equations to physics (Professor Stożek, 4/4).

Theoretical mechanics (Professor Banach, 4/4).

Problem classes in theoretical mechanics (Professor Banach, 2/2).

Differential geometry (Dr. Nikliborc, 3/3).

Problem classes in differential geometry (Dr. Nikliborc, 1/1).

Geometry of convex sets (Dr. Nikliborc, 3/3).

Trigonometric series (Dr. Kaczmarz, 3/3).

Elementary number theory (Reader, Maksymowicz, 0/2).

Logic (Professor Ajdukiewicz, 2/2).

Problem classes in logic (Professor Ajdukiewicz, 2/2).

Theoretical physics (Professor Rubinowicz, 5/5).

Problem classes in theoretical physics (Professor Rubinowicz, 2/2).

Physics seminar (Professor Rubinowicz, 2/2).

The schedule was thus pretty packed (40/39), but S. M. Ulam's industriousness depended on what interested him most. He only sat for his exams under pressure (his last year of regular studies was 1930/1931, but he sat for his last four exams for the course as late as November and December 1932). Almost certainly this is explained by the rapidity with which his mathematical talent matured and by the early identification of his true academic interests, as expressed in an attachment to an application submitted to the Commission for Degree Examinations at the general studies department of Lvov Polytechnic (the application is undated, but it must have been in autumn 1932) listing twelve of his papers, nine of which had already appeared in print (copies were provided), and three of which were pending publication.

To obtain the diploma one had to submit a thesis, in addition to passing the coursework exams, and pass both a written as well as a an oral exam. Ulam's thesis for the diploma was entitled "On the product operation" ["*O operacji produktu*"]. His advisor, K. Kuratowski, assessed it as follows (a direct translation is given):

> This thesis is a study on a subject for which little research has been undertaken to date and which is now playing an increasingly important role in mathematics, namely "taking the product". The author makes an analysis of this notion with reference to set theory, group theory, topology, the geometry of metric spaces, combinatorics, and measure theory associated with probability calculus.
>
> Bearing in mind that the author has demonstrated complete mastery of the subject matter, requiring familiarity with the relevant literature, and that, furthermore, the work contains a series of his own results, and that the author finally poses many interesting questions—I recognize the considered diploma thesis as outstanding.

[Underline as in the original—R. D.]

For the written exam (at the time it was called the "invigilated" exam—"klauzurowy") he had to solve the following four problems (a direct translation of the original text is given here):

1) A bounded, integrable function $f(x)$ of period 2π is given. Consider the sum:

$$S_n = a_0/2 + a_1 \cos x + b_1 \sin x + \cdots + a_n \cos nx + b_n \sin nx.$$

Determine a_i and b_i so that the integral $J = \int_0^{2\pi} (f - S_n)^2 dx$ achieves a minimum.

2) Integrate the equation

$$(mz - ny)p + (nx - lz)q = ly - mx,$$

where l, m, n denote constants and $p = df/dx$, $q = df/dy$.

3) Show that the series $\sum_{n=0}^{\infty} z^n$ cannot be extended analytically outside the circle of convergence[43].

4) Determine the locations of the geometric centers of the surface

$$x^2 + y^2 + z^2 + 2pxz + 2qyz - 2ax - 2by + 2cz = 0.$$

The grade for the closing exam was "very good". Ulam was given ten questions for his oral exam:

(1) Conformal mappings of surfaces.

(2) The Jacobian.

(3) The "strong" law of large numbers.

(4) Definitions and key properties of the product.

[43]This is the original wording. Actually, this series is equal to $z/(1-z)$ for $|z| < 1$, and this function has an analytic continuation over the whole complex plane, except for the point $z = 1$. Probably the series $\Sigma_{n=0}^{\infty} z^{2^n}$ was intended.

(5) The axioms of set theory.

(6) Functions representable analytically and Borel sets.

(7) Analytic sets and projective sets.

(8) In what different ways one can treat analytic functions.

(9) The analogy between Cauchy's integral and certain theorems in the theory of differentiable equations.

(10) Integral equations.

The grade awarded for the oral exam was "outstanding".

In consideration of all this, and mindful of the "outstanding grade obtained in his oral exam for the diploma" in the presence of a commission consisting of W. Stożek (head examiner), A. Łomnicki and K. Kuratowski (members)—the board of the PL general studies department awarded him the "master's degree" on December 15, 1932 (Figure 6).

These were not typical mathematical studies and neither was the final exam. The example above nonetheless provides some insight into the standard of mathematical studies at Lvov.

Mathematical studies at the UJK philosophy department (later the Department of Mathematics and Natural Sciences) were of a similar ilk.

> Course lectures were delivered on a rotational basis by all four professors. There were problem classes for the lectures, for which attendance was compulsory if one wanted to sit for the exams. [...] I sat for only one exam under Professor Steinhaus, namely in theoretical mechanics, but the lectures I listened to were given by Professor Stefan Banach, while I went to Reader Herman Auerbach for the problem classes in mechanics. [...] The assignment of a student to an examiner was made at random. Exam sessions took place after every trimester. Each student declared his/her willingness to do exams in his/her selected subjects by going to the office of the dean, the list of candidates was sent to the mathematical seminars, usually headed by adjunct Auerbach, and here the candidates were divided up, depending on their chosen subjects, between Professors Żyliński, Steinhaus, Banach, and Ruziewicz. The students had to know the tasks required for their programs, and the fact that every examiner had his own special set of tasks is another matter altogether.
>
> [B: Szałajko, 1994; p. 254]

The academic and social life of students revolved to a great extent around the UJK Student Circle for Mathematical Physics.

> Any student of mathematics, physics, or astronomy could join the Mathematical-Physics Circle [of UJK Students]. Upon acceptance, one had to submit a handwritten declaration, ratified

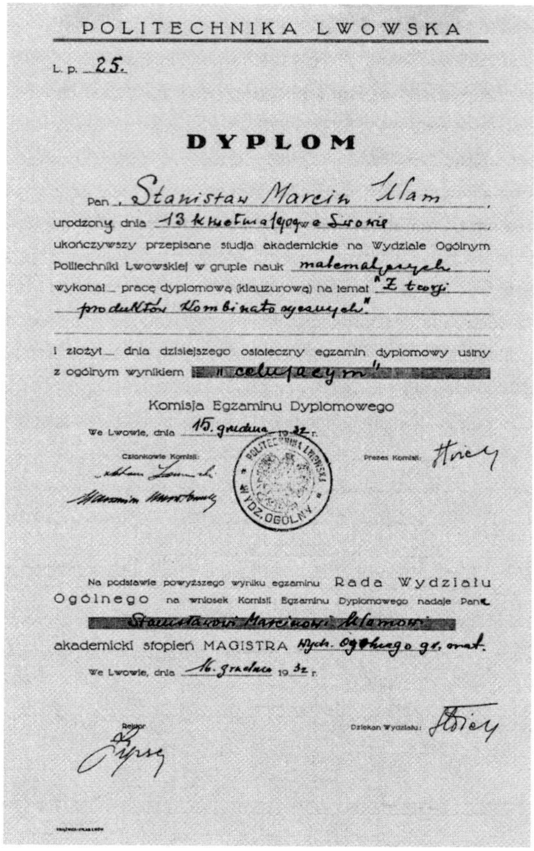

POLITECHNIKA LWOWSKA

L. p. 25.

DYPLOM

Pan Stanisław Marcin Ulam,
urodzony dnia 13 kwietnia 1909 we Lwowie,
ukończywszy przepisane studja akademickie na Wydziale Ogólnym Politechniki Lwowskiej w grupie nauk matematycznych,
wykonał pracę dyplomową (klauzurową) na temat „Z teorji produktów kombinatorycznych"

i złożył dnia dzisiejszego ostateczny egzamin dyplomowy ustny z ogólnym wynikiem „celującym"

Komisja Egzaminu Dyplomowego

We Lwowie, dnia 15. grudnia 1932 r.

Członkowie Komisji:

Prezes Komisji:

Na podstawie powyższego wyniku egzaminu Rada Wydziału Ogólnego na wniosek Komisji Egzaminu Dyplomowego nadaje Panu
Stanisławowi Marcinowi Ulamowi
akademicki stopień MAGISTRA Wydz. Ogólnego gr. mat.

We Lwowie, dnia 16. grudnia 1932 r.

Rektor

Dziekan Wydziału:

FIGURE 6. Stanislaw Ulam's master's diploma.

by two members of the circle. It was not necessary to encourage people to sign up with the circle, because the advantages of membership were obvious and the responsibilities were minimal. Every lunchbreak, i.e. between 1:00 pm and 3:00 pm, [the circle] sat in session in the student reading room [...]. Between those hours the place belonged to the circle. Standing there were several cupboards belonging to the circle. Kept there were its documents and archives, a library and mathematical papers and books for sale. Members of the circle were most interested in the library, which was quite well stocked from the very beginning. It was [...] very useful to students, because it contained the most frequently used basic textbooks, indeed it held multiple copies of them, which could be borrowed and taken home for a period of one month. Therefore the mathematical seminar was well complemented by the library. Of second-most interest to circle members was the editorial commission, which had the task of printing or selling manuscripts, either their own or those of other circles. It also sold antiquarian or commissioned books.

[B: Szałajko, 1984; p. 87]

In the years 1908–1914 the Lvov circle published ten texts by W. Sierpiński (they are listed in [A: Sierpiński, 1974–1976; Vol. I, pp. 61–62]). After the renewal of the circle in 1920, it published, among others, the following texts (titles have been translated):

H. Steinhaus, The Lebesgue Integral, The Mathematical-Physics Circle, Lvov, 1920, 170 pp.

H. Steinhaus, Foundations of Geometry, Mat. Phy. Circle S. U. J. K., Lvov, 1925, 82 pp.

S. Ruziewicz, Selected Topics in Number Theory, The Mathematical-Physics Circle of J. K. University Students, Lvov, 1929, 63 pp.

H. Steinhaus, Fourier Series (edited by J. Schreier and M. Stark), Com. Sen. Acad, UJK, Lvov, 1930, 138 pp.

E. Żyliński, Introduction to Theoretical Arithmetic (edited by M. Stark), Lvov 1932, 38 pp.

In those days all manuscripts were written by hand and printed lithographically. Typewriters were a rarity back then [B: Szałajko, 1984; p. 89].

Testimony to the role played by the circle in the lives of UJK students is seen in the active social life, including "tea parties" ("herbatki"), expeditions, and games.

> Tea parties were organized in a team effort, the culinary side being handled by girlfriends, often with assistance from the wives of the professors. After the official part was over, namely after welcoming the invited guests and listening to the talks, the fun part started and went all night. [...] Tea parties were usually held in the Academic Reading Room situated in the student house located on Łoziński Street (it was popularly known as the "Łoziniec"). Łoziński Street connected Fredro Square with Piłsudski Street. One reached it by going past the Scottish Café. The first tea party took place during the carnival of February 9, 1930. Taking part in it were, among others, Professors Stefan Banach, Eustachy Żyliński, and Włodzimierz Stożek, together with Reader Władysław Nikliborc.
>
> In addition to the parties, which became traditional, the circle made trips to near and more distant places surrounding Lvov. Already in early spring the circle would spend a Sunday or two on an expedition. Typically they would first take the tram to the end of the line and then go on foot. Sometimes they took the train, and sometimes they would even go by horse and cart. The trips were popular, at least at the beginning, with the professors, who did not just go along for the ride but took part in the fun and games organized. Professor Banach went on these trips constantly during the years 1931–1933, the period when he was the curator of the circle. Professor Steinhaus also took part in these trips.

[B: Szałajko, 1984; p. 90]

> Stefan Banach frequently joined these expeditions, was totally relaxed on them and never behaved with even the slightest haughtiness. He also came to the so-called "tea parties" in the student house "at the Łoziniec"; sometimes he came to the carnival games

organized by the circle in the larger ballrooms, for example those at the Post House or the "George" Hotel. And so the circle played a substantial part in the social and academic lives of students of mathematics, physics, and astronomy at the mathematics and natural sciences department of Jan Kazimierz University.

[B: Szałajko, 1993; p. 50]

Zjazd Kół Matematyczno-Fizycznych (Lwów 1930). 1 — L. Chwistek, 2 — S. Banach, 3 — S. Loria, 4 — K. Kuratowski, 5 — S. Kaczmarz, 6 — J. P. Schauder, 7 — M. Stark, 8 — K. Borsuk, 9 — E. Marczewski, 10 — S. Ulam, 11 — A. Zawadzki, 12 — E. Otto, 13 — W. Zonn, 14 — M. Puchalik, 15 — K. Szpunar

FIGURE 7. Convention of the mathematics and physics circle, Lvov, 1930.

The Lvov Student Circle saw many people who became accomplished academics later, after the war. Among them were Andrzej Alexiewicz (Poznań University, professor), Feliks Barański (Cracow Polytechnic, professor), Zdzisław Dąbrowski (Śląsk Polytechnic in Gliwice, senior lecturer), Bogusława Dohnalek-Czyżewska (Częstochowa Polytechnic, senior lecturer), Mark Kac (United States of America, professor), Józef Kiwała (Świętokrzyska Polytechnic in Kielce, reader), Stanisław Leja (United States of America, professor), Józef Meder (Szczecińska Polytechnic, professor), Janina Pokorny-Szałajko (Śląsk Polytechnic in Gliwice, senior lecturer), Jan Przemski (Śląsk Polytechnic in Gliwice, senior lecturer), Kazimierz Szałajko (Śląsk Polytechnic in Gliwice, reader), Adam Wachułka (Pedagogical School in Cracow, reader), Jadwiga Zarzycka-Pawłowska (Cracow Polytechnic, adjunct).

There were similar circles at each of the five prewar Polish universities. These circles worked closely with one another, as can be seen in the yearly conventions that were held on each successive occasion at the home base of a particular circle (Figure 7). The schedules for such conventions always included a general assembly and an academic session with talks given by professors, readers, and assistants.

CHAPTER 8

Journals, Monographs, and Congresses

Lvov hosted the First Polish Mathematical Congress between September 7–10, 1927. Hugo Steinhaus recalled: "The Lvov congress was excellent. I then thought about establishing a new international journal of mathematics in Lvov; I put it to Banach and we began to take the preliminary steps" [B: Steinhaus, 2010; p. 110].

The Lvov journal was based on the same ideas as the Warsaw journal, "Fundamenta Mathematicae" [B: Duda, 1995a], [B: Duda, 1996], which had been appearing since 1920. The new journal was to be dedicated to "operator theory, probability theory, and other fields pursued in Lvov", and not "set theory and its applications", as it was in Warsaw. Even though the scope of the Lvov journal was intended to be different than that covered by "Fundamenta", the proposal was received with reservation in Warsaw. "There was a fear the journal, if it were subsidized by the ministry, would mean a reduction in the grant money available for "Fundamenta" [...]. I maintained the threat was negligible. The ministry awarded us a subsidy and we got to work, but quite a long time elapsed before the first volume appeared" (all quotes from [B: Steinhaus, 2010; pp. 120 and 110]).

Nonetheless, people still had reservations. "The Varsovians insisted we promise not to print papers on set theory: the proviso was superfluous because our program rested on operator theory, probability theory, and other fields pursued in Lvov, but not set theory, which if anything had been afforded too much space in Poland. So it was most surprising when the very person who made us promise to that condition should then send us his own paper on set theory! We didn't include it" [B: Steinhaus, 2010; p. 120].

The new journal was called "Studia Mathematica"[44] (see Figures 8 and 9). It first appeared in 1929, 500 issues being printed, 200 of which were exchanged for others. Of the fourteen papers published in that volume, twelve came from Lvov, their authors being Stefan Banach, Zygmunt Wilhelm Birnbaum, Stefan Kaczmarz, Stanisław Mazur, Władysław Nikliborc, Władysław Orlicz, Juliusz Schauder, Hugo Steinhaus. Over the passing years, nine volumes of "Studia Mathematica" were produced in Lvov (the last of them in 1940[45]), the list of authors extended to Herman Auerbach, Meier Eidelheit, Mark Kac, Stanisław Ruziewicz, Józef Schreier,

[44]Apparently Antoni Łomnicki proposed the name. See [B: Orlicz, 1981; p. 229].

[45]"Studia Mathematica" was revived in Wrocław after the war. The first postwar paper was accepted already in August 1946, but the first Wrocław volume, the tenth in the series, did not appear until 1948. On its jacket cover the names of Stefan Banach and Hugo Steinhaus appeared as its "founders", the listed "editors" were Hugo Steinhaus, Władysław Orlicz, Stanisław Mazur, and Bronisław Knaster, while Wrocław was the stated place of issue. The place of issue became "Warsaw-Wrocław" in Volume 13 (1953) and subsequently, until Volume 101 (1991) when it changed exclusively to "Warsaw". "Studia Mathematica" has been appearing ever since then, and today it continues to remain one of the world's leading journals on functional analysis.

Ludwik Sternbach, Andrzej Turowicz, Stanisław Ulam—and that was the core of the Lvov school of functional analysis.

A total of 161 papers were published in Volumes 1–9 of "Studia Mathematica", which appeared during the years 1929–1940. Of these, 111, in other words about 69%, were authored by Lvov mathematicians, while only 18% of papers came from abroad. Here is a list of Polish mathematicians who contributed at least one paper to Volumes 1–9 of "Studia Mathematica" (the total number of papers contributed by a given author is shown, followed by a roman numeral in brackets, which denotes the volume number of the first paper contributed by that author) [B: Duda, 1995a]:

W. Orlicz	21 (I)	W. Nikliborc	5 (I)
S. Mazur	17 (I)	Z.W. Birnbaum	4(I)
S. Banach	16 (I)	Z. Zalcwasser	3(II)
S. Kaczmarz	12 (I)	S. Saks	2 (I)
H. Steinhaus	9 (I)	W. Sierpiński	2 (II)
H. Auerbach	9 (II)	L. Sternbach	2 (IV)
M. Kac	9 (V)	S. Ulam	2 (IV)
J. Marcinkiewicz	8 (VI)	S. Ruziewicz	1 (III)
M. Eidelheit	7 (VI)	A. Turowicz	1 (VIII)
J. Schauder	7 (I)	D. Wajnstein	1 (IX)
J. Schreier	6 (II)	M. Wojdysławski	1 (IX)
A. Zygmund	6 (II)		

Of the fourteen Polish authors who published at least four papers, only J. Marcinkiewicz and A. Zygmund did not come from Lvov.

The next important event for the Lvov mathematical community was the appearance of Banach's monograph [A: Banach, 1932], constituting a synthesis of achievements in the theory of linear spaces equipped with norm and containing results of its author and of the Lvov school (see Chapter 20). It was an expanded edition of an earlier book that had appeared in Polish ([A: Banach, 1929b], [A: Banach, 1931c]); S. Mazur wrote the *Remarques* and B. Knaster translated it into French. Clarity of style, powerful results, and a multitude of valuable examples ensured the monograph quickly acquired wide popularity; it marked the moment when the Lvov School of Mathematics began to be recognized more widely (see Chapter 21).

We add here that a second monograph written by Lvov mathematicians appeared, namely [A: Kaczmarz & Steinhaus, 1936], in the monograph series Monografie Matematyczne, which had started off so well with Banach's book. It summarized its authors' long-standing interest in trigonometrical series and, more generally, orthogonal series (H. Steinhaus himself produced twenty papers on the subject, including a joint paper with S. Kaczmarz). There appeared two more monographs by authors connected with Lvov, namely [A: Kuratowski, 1933] and [C: Zygmund, 1935]: the first of these was an original and important monograph on set topology, later complemented with Volume II, which had many new additions, including those in Russian and English; the latter work became the foundation stone for the school which its author would later set up in the United States.

The three mathematical congresses that took place in 1928 (Bologna), 1932 (Zurich), and 1936 (Oslo) provide an interesting illustration of the attitude towards mathematics as practiced in Lvov. When the first of these congresses took place,

STUDIA MATHEMATICA

REDAKTOROWIE

STEFAN BANACH HUGO STEINHAUS

TOM I.

LWÓW WYDANO Z SUBWENCJI MINISTERSTWA W. R. i O. P. 1929

FIGURE 8. Title page of the first volume of "Studia Mathematica".

Banach had already published his doctoral thesis, as well as a few other papers, but "Studia Mathematica" had not yet appeared (Volume I was still being prepared) and the great talents of S. Mazur, W. Orlicz, M. Kac, S. Ulam, and others were not given the opportunity to shine. It is therefore no surprise that nobody talked about the results of S. Banach or of H. Hahn or of N. Wiener, even though the plenary speakers included such luminaries of functional analysis as J. Hadamard, V. Volterra, and M. Fréchet[46] (J. Hadamard noted only that S. Banach and N. Wiener had given a "deeper" formulation of function spaces, while M. Fréchet did actually mention the names of H. Hahn and F. Riesz, but not that of S. Banach, as an example of international collaboration on work at the boundary of analysis and set theory.) This was not a symptom of any ill will. It was simply the fact that the work flowing from Lvov had not yet been fully appreciated. It is true that H. Steinhaus and S. Kaczmarz, both of them from Lvov, delivered sessional talks, reporting on progress that stemmed from S. Banach's concepts, including some new, elegant, and powerful results ([A: Steinhaus, 1930b], [A: Kaczmarz, 1930]), but this still did not alter the general picture. "Most surprisingly, the Bologna Proceedings gave no sense of the creative ferment generated in Lvov by the systematic study of Banach

[46][C: Hadamard, 1929], [C: Volterra, 1929], [C: Fréchet, 1929].

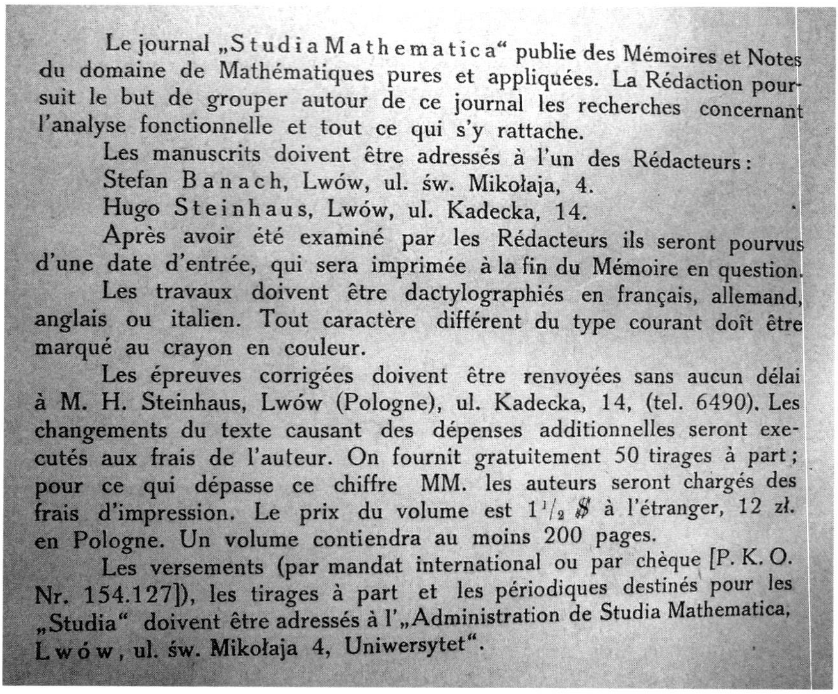

Le journal „Studia Mathematica" publie des Mémoires et Notes du domaine de Mathématiques pures et appliquées. La Rédaction poursuit le but de grouper autour de ce journal les recherches concernant l'analyse fonctionnelle et tout ce qui s'y rattache.

Les manuscrits doivent être adressés à l'un des Rédacteurs:
Stefan Banach, Lwów, ul. św. Mikołaja, 4.
Hugo Steinhaus, Lwów, ul. Kadecka, 14.

Après avoir été examiné par les Rédacteurs ils seront pourvus d'une date d'entrée, qui sera imprimée à la fin du Mémoire en question.

Les travaux doivent être dactylographiés en français, allemand, anglais ou italien. Tout caractère différent du type courant doît être marqué au crayon en couleur.

Les épreuves corrigées doivent être renvoyées sans aucun délai à M. H. Steinhaus, Lwów (Pologne), ul. Kadecka, 14, (tel. 6490). Les changements du texte causant des dépenses additionnelles seront exécutés aux frais de l'auteur. On fournit gratuitement 50 tirages à part; pour ce qui dépasse ce chiffre MM. les auteurs seront chargés des frais d'impression. Le prix du volume est $1^1/_2$ $ à l'étranger, 12 zł. en Pologne. Un volume contiendra au moins 200 pages.

Les versements (par mandat international ou par chèque [P. K. O. Nr. 154.127]), les tirages à part et les périodiques destinés pour les „Studia" doivent être adressés à l'„Administration de Studia Mathematica, Lwów, ul. św. Mikołaja 4, Uniwersytet".

FIGURE 9. The inside back cover of the first volume of "Studia Mathematica".

spaces, or of the stimulus already being provided by the foundations of quantum mechanics. How misleading this turned out to be!" ([B: Birkhoff & Kreyszig, 1984; p. 306]).

The situation was similar four years later at the next congress, in Zurich. A large twenty-strong group of Polish mathematicians participated in it. Those attending from Lvov included K. Kuratowski (then a professor at the general studies department at Lvov Politechnic) and S. Ulam. W. Sierpiński gave a plenary address, and a further ten sessional talks were given, two of which were given by Lvov mathematicians, but once again there was total silence on the subject of Banach spaces and the achievements of the Lvov school. However, it was the year when S. Banach's monograph appeared [A: Banach, 1932] and by which time three earlier volumes of "Studia Mathematica" had already been published.

The importance of S. Banach's monograph can be seen from the fact that at the following congress (Oslo, 1936), the situation looked entirely different. On that occasion S. Banach was not only present, he also gave a plenary talk [A: Banach, 1937] in which he presented the history of "operator theory" (seen from his point of view) and results obtained in Lvov regarding "type (B) spaces" (S. Banach kept using that name, though type (B) spaces were increasingly being called Banach spaces).

For at least thirty years after its first appearance, S. Banach's monograph remained a fundamental source of reference and information. That is enough time for the terminology to be adopted and applied and also for a particular way of looking at the theory to be established. Consequently, even though its role as a source text was superseded in the 1960s by another massive monograph [C: Dunford

& Schwartz, 1958] covering much work done right after Banach's death and after the liquidation of the Lvov school, it still remains the theory of Banach spaces, and his name is one of the most frequently quoted in mathematical literature around the world. He is without doubt the world's best known Polish mathematician.

CHAPTER 9

The Popularization of Mathematics

The small number of salaried positions and the lOw pay at UJK and PL forced active mathematicians to take up extra work, most often in schools. Juliusz Schauder was one of them. Years later his former pupil Alfred Jahn[47] recalled what kind of a teacher he was:

> I should mention Juliusz Schauder in particular. I remember how one day he appeared in our class. A young, slightly balding person with a determined expression on his face, he declared he was going to teach us mathematics. We shortly learned he was a reader at the university, was an outstanding academic, and was teaching at the school in order to earn a little extra money, because the university pay was so low. [...] We thought—and we pretty much counted on it—that as an academic at the university he was not going to take school teaching too seriously. We were happy at the prospect of having an easy time. Then something unusual happened. Schauder treated us like nobody else had done before. The year was 1929 and I was in the fourth year. His teaching was excellent. Lessons in a difficult subject, and generally not much enjoyed, became unusually attractive to us. He was extremely demanding, but he had the right to be, because he could teach mathematics to anyone who at least had a minimal commitment and interest in it. [...] He shared his uncommon passion for the subject, and delivered his lessons with a special kind of vitality. [...] Schauder didn't merely sit behind a desk, he walked around the room, prepared to explain things individually. He could motivate some of the boneheads at the back of the class; lessons were too short for him, so they overran into break time. At last, when he noticed a group of more able and more motivated students was forming, he proposed they all meet at the school in the afternoons. Amazingly, we came because we wanted to, he didn't get paid anything from anyone for doing this, and nobody forced us to do this. [...] Schauder taught us for two years.

[B: Jahn, 1991; p. 33]

[47]Alfred Jahn was an alumnus of Jan Kazimierz University. He left Lvov in 1945 and lived in Wrocław since 1945. As one of the world's leading authorities on polar geomorphology, he participated in may expeditions to Greenland, Spitsbergen, Alaska, and Siberia. Between 1962–1968 he was the rector of Wrocław University.

The joy of doing mathematics led to the joy of its popularization. There were therefore talks for the general public and one of those talks made a big impression on S. Ulam [B: Ulam, 1969].

Many of the mathematicians working at UJK and PL were involved with writing academic textbooks and school textbooks. We list below (in chronological order) some of the more memorable works—given here are only translations of the original titles, their years of publication, and sometimes the number of pages:

K. Bartel, *Descriptive geometry*, 1919 and 1922. Updated after the war, in 1948 and 1958.

S. Ruziewicz and E. Żyliński, *Introduction to mathematics. Elements of higher algebra and number theory*, 1929, vii + 269 pp.

S. Banach, *Calculus*, Vol I, 1929, 294 pp.

S. Banach, *Calculus*, Vol II, 1930, 248 pp.; both of Banach's volumes were updated several times, starting with a reproduction in 1949 and ending with the corrected editions (4th edition in 1952, 5th edition in 1955, 6th edition in 1957), together with a translation into Russian, done by S. I. Zuchovicki, 2nd edition in Moscow in 1966, 3rd edition in Moscow in 1972.

W. Nikliborc and H. Steinhaus, *Exercises in calculus*, 1930, 260 pp.

W. Stożek, *Lectures in mathematics I*, Part I, with worked solutions by W. Nikliborc, 1934.

W. Stożek, *Lectures in mathematics I*, Part II, with worked solutions by W. Nikliborc, 1934.

A. Łomnicki, *Calculus for natural scientists and engineers*, Vol I, 1935.

A. Łomnicki, *Calculus for natural scientists and engineers*, Vol II, 1936.

S. Banach, *Mechanics for high schools*, Part I, 1938, v + 234 pp.

E. Żyliński, *Analytic geometry*, 1938, vi + 386 pp.

A. Łomnicki, *The Platonic solids*, 1939.

A. Łomnicki, *Infinite series, differential equations, differential geometry*, 1939.

A. Łomnicki, *Differentiable and integral calculus for natural scientists and technicians* (updated, after 1945).

S. Banach, *Introduction to the theory of functions of a real variable*, 1951.

Independently of their inherent merit, the textbooks were also valuable because back then there were still very few Polish academic textbooks in mathematics (written in Polish, by Polish authors). Students had to use books printed in foreign languages, which were expensive and very difficult to obtain, or else they had to depend on lecture notes.

Some Lvov mathematicians wrote school textbooks. Among them are the following (listed in chronological order according to the year of the first edition). Listed below are translations of the original Polish or Ukrainian titles:

A. Łomnicki, *Geometry* [school textbook], Part I: Plane Geometry and Solid Geometry, 1911 (updated in separate Parts I and II, 1914, 1920, 1923).

A. Łomnicki, *Geometry*, Parts III and IV: Trigonometry, Analytic Geometry [school textbook], 1920 (four editions up to 1930).

A. Łomnicki, *Elementary geometry* [school textbook], 1920 (six editions up to 1929).

A. Łomnicki, *The foundations of analytic geometry*, 1921 (three editions up to 1923).

A. Łomnicki, *Four-digit mathematics-physics tables*, 1921 (thirteen editions up to 1951).

S. Ruziewicz, and E. Żyliński, *Algebra*, upper level textbook for schools, three parts, 1926–1928, 196 + 142 pp.

S. Banach, W. Sierpiński, and W. Stożek, *Arithmetic and geometry for high school year I*, 1929.

S. Banach, W. Sierpiński, and W. Stożek, *Arithmetic and geometry for high school year II*, 1930.

S. Banach, W. Sierpiński, and W. Stożek, *Arithmetic and geometry for high school year III*, 1931.

S. Banach, W. Sierpiński, and W. Stożek, *Arithmetic and geometry for common schools year V*, 1933.

S. Banach, W. Sierpiński, and W. Stożek, *Arithmetic and geometry for common schools year VI*, 1933.

S. Banach, W. Sierpiński, and W. Stożek, *Arithmetic for lower school year I*, 1933 (updated 1937).

S. Banach, W. Sierpiński, and W. Stożek, *Arithmetic for lower school year II*, 1934 (updated 1939).

S. Banach, W. Sierpiński, and W. Stożek, *Arithmetic for lower school year III*, 1935 and 1947.

S. Banach, W. Sierpiński, and W. Stożek, *Arithmetic and geometry for common school year VII*, 1935.

A. Łomnicki, *Geometry for lower school year III*, 1935, 128 pp.

W. Nikliborc and W. Stożek, *Algebra for lower school year III*, 1935, 128 pp.

S. Banach, *Algebra for lower school year IV*, 1936 (updated: Jerusalem, 1944; Bari, 1946; Edinburgh, 1946; Wrocław 1946).

A. Łomnicki, *Geometry for lower school year IV*, 1936.

W. Nikliborc and W. Stożek, *Algebra for lower school year IV*, 1936, 112 pp.

S. Kubraczkiewicz and W. Orlicz, *Calculations for common school year I*, 1937.

W. Nikliborc and W. Stożek, *Arithmetic and geometry for common school year IV*, level I, 1937, 128 pp.

A. Frejlich and W. Orlicz, *Mathematics for secondary common schools*, 1938 [in Ukrainian].

A. Łomnicki, *Algebra for year I classes of general education secondary schools*: humanities department, 1938.

A. Łomnicki, *Trigonometry*, textbook for general education secondary schools, 1938.

W. Nikliborc and W. Stożek, *Arithmetic and geometry for common school year IV*, level I, Course A, 1938, 128 pp.

W. Nikliborc and W. Stożek, *Arithmetic and geometry for common school year IV*, level I, Course C, 1938, 144 pp.

A. Łomnicki, *Mathematics for high schools specializing in mechanics, electrical systems, roads, water-treatment and telecommunications: Analytic geometry, calculus*, 1939 (updated 1947).

A. Łomnicki, *Mathematics for all-male high schools for industry: Algebra and trigonometry*, 1947.

H. Steinhaus recalled the reason why S. Banach was interested in writing school textbooks:

> Banach could work anywhere at any time. He was not accustomed to comforts and he did not need them, so a professor's salary should have been sufficient for him. But a love of the café scene, a total lack of home savings, and a failure to budget for everyday matters got him into debt, and ultimately into a very difficult situation. Wanting to extricate himself, he took up writing textbooks.
>
> [B: Steinhaus, 1960; p. 156]

A contemporary assistant recalled how he wrote those textbooks:

> To understand the way Banach wrote school textbooks, it is worth describing, as an illustrative example, the preparation of a textbook for year IV high school classes. I myself was actively involved with the preparation. It happened in June 1936. I was at the time a young assistant at the same chair as Professor Banach. To my astonishment he asked me to help him prepare that textbook for publication. My help only actually lasted a few days, but it was a very intensive time. My job was to correct the typewritten copy of his handwritten manuscript, to sort all the drawings and stick them in. I also had to liaise between the author and Książnica-Atlas, a joint-stock company that was financing the textbook and which had guaranteed royalties to the author and which was to publish the aforementioned textbook, pending approval by a special ministerial commission. Eight typewritten copies had to be delivered to the ministry by 8 a.m. on Monday, June 1. Meanwhile, not quite one week before the deadline, hardly a few pages had been typed up. That's when the furiously hard work started. It went on day and night without a break, from the Thursday to the Sunday before the deadline. Naturally the professor, Reader Auerbach, a draughtsman, and myself were all there. And we met the deadline too: before Sunday afternoon I took the last pages of the manuscript to the publisher. I think the entire staff of Książnica were on standby: not just the typists, but also the bookbinders, the people who did calculations, the bookkeepers, and others still. The last typewritten pages were ready around three o'clock in the afternoon. I returned to the professor and told him the manuscript had been delivered. We all sighed in relief. I was really pleased with myself because I didn't mess anything up during those few crazy days, and equally I had the great satisfaction of working with Professor Stefan Banach. I occasionally stopped by the Scottish Café, where the professor invited me for coffee, to help me keep my strength up. In the end I earned almost three times my monthly salary, paid for by the publisher on the personally written instructions of the professor.
>
> [B: Szałajko: 1994; p. 255]

H. Steinhaus's passion for popularizing mathematics was revealed around this time, when he started writing popular booklets and articles, e.g. [A: Steinhaus, 1923b] and [A: Steinhaus, 1927]. However, he deservedly received the greatest recognition for *Mathematical Snapshots*, an unusual book that fused the powers of observation of a researcher—seeing how mathematics underlies everything from chess games to nature—with the talents of a popularizer [A: Steinhaus, 1938b]. Starting with its first Polish edition in 1938, there were three more editions in Poland (the last was in 1989). It has also been translated into several dozen languages, including English (four editions), Russian, Hungarian, Czech, Japanese, German, Rumanian, French, Bulgarian, and others. Of all the books on mathematics ever written by a Polish author, it is the one best known internationally.

CHAPTER 10

Social Life (the Scottish Café, the *Scottish Book*)

Lvov liked to have fun and social life thrived.

> The dance craze, excursions, libertine excesses, betrayals, and short-term relationships, in other words enjoyment of freedoms dreamed of for years: helping oneself to rewards after the pressures and dark days of the war years—all of this was typical of the years 1920–1923.
>
> [B: Steinhaus, 2010; p. 99]

As the "dark days of war" receded from memory, the prime order of the day was a desire for success, to prove to ourselves that Poles deserved a free state, that they were capable of uniting ourselves after three partitions and to build anew, and that, most particularly, Polish learning (Lvov included) could "strike out for independence". And the attainment of that freedom was followed with remarkable successes—an enduring joy of life. A contemporary student reminisces on the period 1925–1929:

> I was someone who belonged to that generation. Mathematics for those fascinated by it was something akin to a burning fever. We went together to all manner of places at various times of the day or night, talking endlessly about mathematics. In a place, which was a cross between a room and small library, there stood a big, old, tiled stove, one side against the wall. I remember having long discussions when it was freezing outside, when we would glue ourselves against the tiles of the three remaining sides, talking about mathematics around the corners of the stove.
>
> (Z. W. Birnbaum, quoted from [B: Woyczyński, 1997; p. 140])

A singular joie de vivre and a fascination with mathematics were manifest in the vibrant café culture.

> A significant part of our mathematical discussions took place in cafés neighboring the university. The first of them was called "Roma". After a year, maybe two, Banach decided to move our sessions to the "Scottish Café", situated on the other side of the street. [...] The tops of the café tables were marble slabs on which one could write in pencil and were, even more importantly, quickly erasable. In our mathematical discussions a single word or gesture, requiring no further explanation, often sufficed to understand what was being meant. Sometimes an entire discussion consisted of just a few uttered words punctuating long periods

of thought. An observer sitting at another table could overhear short bursts of conversation and see lines being scribbled on the table, sometimes interspersed with a laugh from one of the people seated, followed by long periods of silence, during which we consumed coffee and looked vacantly at each other. The acquired habit of persistence and concentration, sometimes lasting hours, was for us one of the most important elements of mathematical work.

[B: Ulam, 1969, p. 52]

FIGURE 10. View of Fredro Square, as seen from St. Nicholas Street. Straight ahead is Łoziński Street, where the "Łoziniec" was located, the Scottish Café was on the right, the Roma Café was on the left.

The table used by Banach, Mazur, and Ulam was the most important one in the Scottish Café. One session there even lasted seventeen hours—resulting in the proof of an important theorem on Banach spaces—but nobody wrote it down, and even today nobody has recovered it [...]. A table cover with traces of pencil lead was probably all that remained of that session and was most likely scrubbed off by the café cleaner. So therefore a great service was done by Łucja Banach, who now rests in a cemetery in Wrocław—when she purchased a thick exercise book with a thick cover and entrusted it to paying customers of the Scottish Café—that was where mathematical problems were written down, on the first side of consecutive pages so that solutions could be added in later, next to the problem statements. The original "Scottish Book" was available to every mathematician

> who asked for it at the café: some problems promised the reward of a prize for a solution—the prizes ranged from a small cup of coffee to a live goose.
>
> [B: Steinhaus, 1961a; p. 257]

The exercise book purchased by Mrs. Łucja Banach, the wife of Stefan Banach, quickly became known as the *Scottish Book* (and is cited as *"The Scottish Book"* in the literature). In the few years of its existence it played a big role in the life of Lvov mathematicians. The first problem in the *Scottish Book* was written in by Stefan Banach on July 17, 1935, the last by Hugo Steinhaus on May 31, 1941. All together it includes 193 numbered problems, some with solutions, encompassing such fields as summability theory, functional analysis, classical analysis, group theory, measure theory, set theory, probability theory, and topology.

FIGURE 11. Inside the Scottish Café during its heyday.

The *Scottish Book* was thus in use for almost exactly six years, from July 17, 1935, to May 31, 1941. The first year was the best because in the summer and autumn of 1935 almost half the problems were entered into it. The precise number is difficult to state because not every author included the date. In any case, the last problems dated for 1935 have the numbers 109 and 110 (October 1935), while the next problem bearing a date is problem number 132 and is dated as late as February 25, 1936. So certainly 110 problems were entered into the *Scottish Book* in 1935, and almost certainly so were a few more, possibly even a dozen or so more. Such a large number of problems testifies to the newly found popularity of the *Scottish Book*, though many of them had arisen in earlier years and had not been solved. The authors of problems 1–110 are exclusively all Lvov mathematicians (in the order of the entries, if there are several authors, we note all their names): Stefan Banach, Stanisław Ulam, Józef Schreier, Stanisław Mazur, Władysław Orlicz, Herman Auerbach, Juliusz Schauder, Hugo Steinhaus, Stanisław Ruziewicz, Zbigniew Łomnicki, Kazimierz Kuratowski.

Another forty-five problems were added between October 1935 and the end of 1936 (the last entry, dated November 18, 1936, was number 155) and only then are there entries from other Lvov mathematicians and guests from outside Lvov. In the remaining calendar years 1939–1941 only thirty-eight problems were written in and they were distributed almost equally (see Figure 13), though in the last

FIGURE 12. S. Mazur and S. Ulam on a street in Lvov, sometime around 1935.

years of peacetime there was a rising number of guest contributors (A. J. Ward, S. Stoïlov, J. von Neumann, Edward Szpilrajn (Marczewski), Karol Borsuk, C. Offord, M. Kampe de Feriet, and others). During this time Meier Eidelheit made his contribution, entering five problems. It is curious that even after the Soviets took over Lvov, the *Scottish Book* continued to be in use, because in that period the last twelve problems were written in, principally by those who had fled Warsaw (B. Knaster, S. Saks, E. Szpilrajn (Marczewski) and Soviet mathematicians (N. N. Bogolyubov, P. Aleksandrov, S. Soboloev, A. P. Bermant, L. Lusternik)). The only locals to write in their problems were S. Banach and, lastly, Hugo Steinhaus.

Although the last problem in *The Book* bore the number 193, there were in fact many more entries. Ignoring the fact that some problems actually consisted of many others, there were also unnumbered entries, specifically occurring after

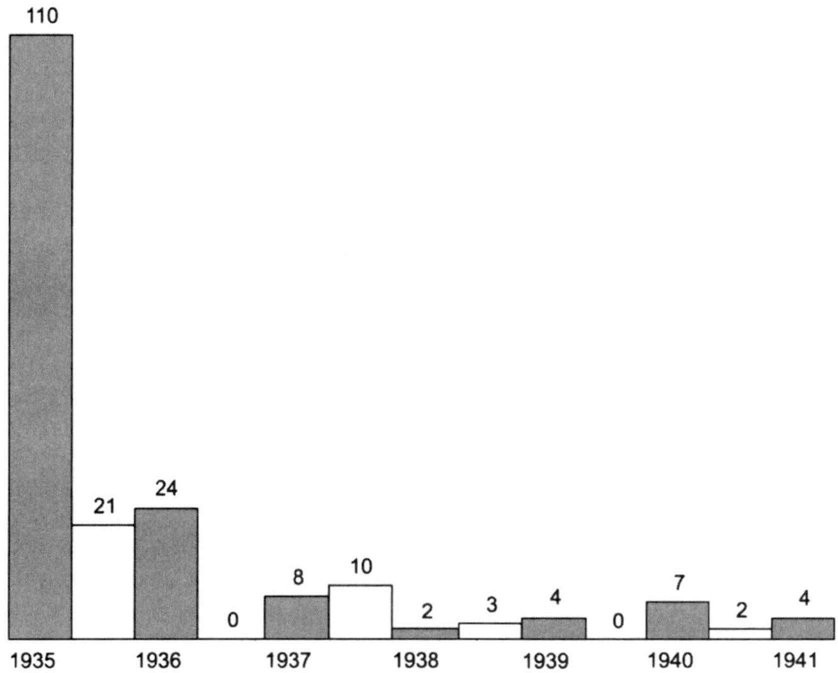

FIGURE 13. The *Scottish Book* in numbers. The chart shows the number of problems entered into the *Scottish Book* over successive years. Since there is no date for many of the entries, the shaded columns denote unambiguous data, whereas the unshaded columns denote the number of problems entered between the last dated entry of the previous year and the first dated entry of the subsequent year.

numbers 10, 15, 17, 20, and 188 (in the monograph [B: Mauldin, 1981] they are labelled 10.1, 15.1, etc.). In total there were thus 198 problems and the statistics below will refer to that number. S. Banach wrote fourteen of his own problems and another nine with others: four with Mazur, four with S. Ulam, and one with both of them. S. Mazur had more, since he wrote twenty-two problems of his own and another twenty-five with other mathematicians, including W. Orlicz (twelve) and S. Ulam (seven). The record holder, in terms of the number of entries, was S. Ulam, who made forty independent entries and twenty-two joint entries, including Banach (five), J. Schreier (six), and S. Mazur (seven). W. Orlicz made three independent entries and twelve joint entries with S. Mazur. H. Steinhaus made nine independent entries. Approximately five to ten further problems were written in by J. Schreier (four independently, six with Ulam), H. Auerbach (four independently and four together with Banach, S. Mazur, or S. Ulam), S. Ruziewicz (five independently), M. Eidelheit (five independently), S. Eilenberg (a series of six problems, written in a single day). The statistics reveal those who were both active mathematicians as well as frequent visitors of the Scottish Café. It also reveals who worked closely with whom; for example Banach's mathematical partners were S. Mazur and S. Ulam (the academic paths pursued by S. Banach and H. Steinhaus had diverged by that time). Some of the problems were of an elementary nature, such as problem number 59, of S. Ruziewicz, asking whether one can break up a square into finitely many different squares, and problem number 123, known as the Steinhaus "sandwich" problem. However, there were also some problems which led to many years of research, such as problem number 43 of Mazur with the definition of "Mazur's game"; problem number 54, of Schauder, about fixed points; problem number 77 of Ulam (if A^2 and B^2 are homeomorphic (isometric, isomorphic) does it follow that A and B are homeomorphic (isometric, isomorphic)?); problem number 153 of Mazur, with the promised reward of a live goose for the solution (see Chapter 21); or problem number 175 of Borsuk (is the product of a triod—i.e. a space homeomorphic to the letter "T"—and a Hilbert cube also a Hilbert cube?). Just about every problem entered into the *Scottish Book* can be said to have inspired interesting ideas and led to interesting and important research. Some questions in it remain open to this day.

After the war and after the death of her husband, Łucja Banach and her son Stefan (who later became a medical doctor) were "repatriated" (meaning compulsory resettlement in Poland, within its new borders) to Wrocław. Knowing the importance of the *Scottish Book*—she took it with her. While in Wrocław and right up until her death after the war (she died in 1954, in Wrocław), she kept in contact with Wrocław mathematicians, above all with Hugo Steinhaus.

> After the war Steinhaus wrote out by hand (word for word!) the contents of the book and in 1956 he sent a copy[48] to Stanisław Ulam in Los Alamos. Ulam transcribed it into English and then had 300 copies made (at his own expense), sending them to friends and various universities around the world. The book became famous and many mathematicians requested Ulam to send further copies. After a few years there were so many requests that it was decided, in Los Alamos, to issue another edition (together with various corrections), but no longer at Ulam's own expense, and which was realized in 1977. In May 1979 at North Texas University a "Scottish Book Conference" took place (Ulam, Kac, and Zygmund were among the participants). Together with updated material regarding the solutions of problems and other problems on associated themes, a book (with additional papers from the conference, most particularly including reminiscences) was published by Birkhäuser in 1981 (edited by Daniel Mauldin).
>
> [B: Ciesielski, 1993a; p. 66], [B: Mauldin, 1981]

Today the original *Scottish Book* is in the custody of the Banach family, and a copy of the original is in the Library of the Mathematical Institute of the Polish Academy of Sciences in Warsaw. Of the various versions and updates, the most worthy is the monograph [B: Mauldin, 1981], referred to above by K. Ciesielski and including much interesting information about every single problem in the *Scottish Book*.

> For those of us who were brought up in the golden age of functional analysis, the *Scottish Book* was and remains a romantic

[48]It was a "typewritten copy"; see [B: Maudlin, 1981; p. xii].

> source of our mathematics [...]. The astonishing problems in the *Scottish Book* expressed the spirit of contemporary mathematics.
>
> (G.-C. Rota, quoted in [B: Mauldin, 1981; dust cover]).

But let us return to mathematics.

> Mathematics evolved ever more vigorously in Lvov. [...] Specialists emerged in different fields. Mazur worked on linear methods in summability, Auerbach on differential geometry, Schauder on partial differential equations, Kaczmarz and Orlicz on orthogonal series, Nikliborc on celestial mechanics and balanced shapes—to mention but a few.
>
> The most interesting aspect was the joint work. For example, the work on the principle of condensation of singularities was written jointly by myself and Banach. The work on the decomposition of a ball into parts, from which one can assemble a larger ball, was undertaken by Banach and Tarski. Mazur and Orlicz wrote joint papers, as did the younger Łomnicki with Ulam, Ruziewicz with Sierpiński, etc. If one were to link together all the names of people who wrote joint papers, one could find paths of names from Polish mathematicians to mathematicians in the world's major countries.
>
> One could identify the different "schools" as well. Banach's school focused on operator theory, Kuratowski's school focused on topology, Chwistek's on mathematical logic: I tried to steer my students towards probability theory and applied mathematics. At the Polytechnic, Bartel worked on the relationship between descriptive geometry and the spatial representation of art; accordingly, he studied Italian paintings and engravings and produced a beautiful treatise on them.
>
> [B: Steinhaus, 2010; pp. 120–122]

K. Kuratowski bears interesting testimony about life in Lvov at the time:

> Taking up the chair in Lvov, I kept my readership position in Warsaw (taking an annual leave as a reader), because I wasn't sure if I could bear to live away from my home city of Warsaw.
>
> It turned out differently: after a year I resigned from my readership position in Warsaw and became fully enticed by Lvov.
>
> How did that happen? The uncommon beauty of the city, which I remember even now with a certain emotion, and the academic way of life, which absorbed me with lightning speed. In particular that part of the academic environment in which I could work closely with others, that is the mathematical environment, was especially appealing to me. Above all, Banach and Steinhaus were there [...].
>
> This Lvovian "climate" was equally conducive to my creative output, so that the years I spent in Lvov were the most active period in my academic life.
>
> [B: Kuratowski, 1981; pp. 86, 89]

A characteristic feature of this academic climate was a fondness for jokes. Once H. Auerbach bought himself a new hat. Shortly afterwards somebody switched it in the Scottish Café for an old and dirty one. Auerbach began wearing it, and when asked why he never cleaned it, he answered that he was not going to clean a hat on behalf of a thief.

One time L. Chwistek came to a lecture wearing one brown shoe and one black one. One of the students noticed this. Chwistek was embarrassed: "It's strange. At home I have another pair just like it."

One of the outstanding Lvov mathematicians was referring to something in a chaotic manner. Somebody caustically observed: "You, sir, are thinking about A, say B, write C, but it should be D."

CHAPTER 11

The Polish Mathematical Society

The start of organized mathematical life in Lvov harks back to 1917, when Zygmunt Janiszewski established a mathematical society there. Its members included Stefan Banach, Zygmunt Janiszewski, Zdzisław Krygowski, Antoni Łomnicki, Stanisław Ruziewicz, Hugo Steinhaus [B: Steinhaus, 2010; p. 99], and Józef Puzyna was elected as its head. However, circumstances at the time were not suitable for the mathematical society to function: Lvov was a city situated on the war front (World War I, the Ukrainian-Polish War, the Russo-Polish War), the head of the society, J. Puzyna, fell ill and died in June 1919, and its founder, Zygmunt Janiszewski, died eighteen months later.

When the Cracow Mathematical Society was established on April 2, 1919, it attracted mathematicians from outside Cracow (Stefan Banach was one of the founders, and Antoni Łomnicki, Hugo Steinhaus, and Eustachy Żyliński were among the first members from Lvov). The thought then came to mind to establish the Polish Mathematical Society. At a general meeting in Cracow, on April 21, 1920, the Polish Mathematical Society (PTM) came into being. The statutes of the PTM still did not, however, allow for the establishment of regional branches, but that was changed on December 22, 1920. Soon afterwards, at the beginning of 1921, the Lvov Mathematical Society became the Lvov branch of the PTM, and it was the first PTM branch to be situated outside Cracow (the branches in Warsaw, Poznań, and Vilnius were all set up in 1923).

Not a great deal is known about the earliest years of the Lvov PTM branch. It was certainly a rather inactive branch because life (not just mathematical life) quite generally was still in the process of recovering after three wars.

Between 1924–1939 the number of members of the Lvov branch of the PTM fluctuated between thirty to forty-two people (for example in 1938 there were thirty-four members). It became more active in 1927, when it organized the First Polish Mathematical Congress, held in Lvov between September 7–10, 1927. There were approximately 200 participants, five of whom came from abroad [B: Księga Pamiątkowa, 1929]. In the words of Steinhaus, "The congress was excellent" [B: Steinhaus, 2010; p. 110]. After 1927 there appeared regular sessional reports from the Lvov center, published as part of *Comptes Rendus*, in the journal "Annales de la Société Polonaise de Mathématique" [49]. They span the period November 1927–October 1938, and from the dates of the sessions one can identify the principal speakers and the titles of their talks. In some cases abstracts of the talks are provided. Since they were written by various authors (first in German, and then in French, after 1935), they are uneven; some of the abstracts are long and detailed whereas in other cases they were frequently absent. It is nonetheless fascinating

[49] See volumes 9 (1931), 10 (1932), 12 (1934), 13 (1935), 15 (1937), 16 (1938), and 17 (1938).

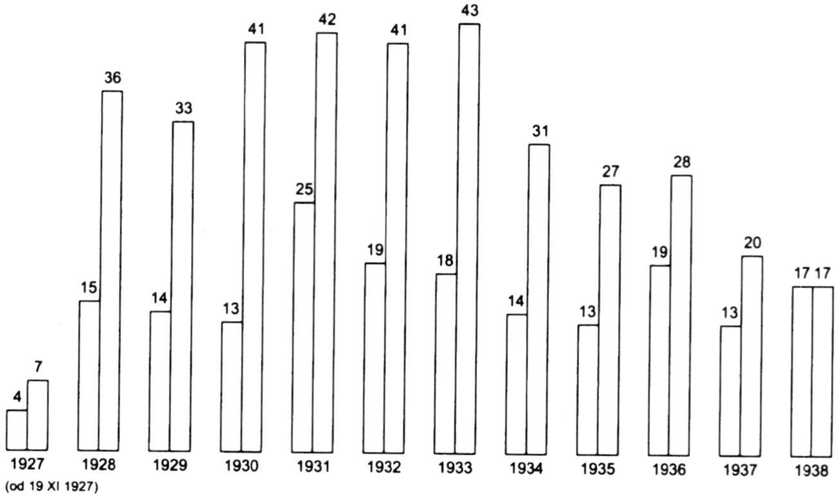

FIGURE 14. Lvov branch of the PTM in numbers. The left column of each column-pair shows the number of sessions in the given year—the right one shows the number of communications given during those sessions.

material, providing interesting insight into how the Lvov branch of mathematics was run.

Let us begin with some statistical data. In the period during which the reports were made, i.e. in the years 1928–1938 (omitting the two months in 1927), a total of 180 sessions were held, during which 360 talks were given[50], which averages out to sixteen annual sessions with two talks per session. There were sixty-five speakers (if a talk was based on the work of several authors, all the contributing authors are counted as speakers). In nearly every case the speakers were active Polish mathematicians. There was a total of sixteen guest speakers who came from abroad.

Moving on now to a more thorough analysis, we note the variation in the number of sessions and the number of talks given (Figure 14). As one can see, the most active period of the Lvov branch was the beginning of the 1930s, specifically the years 1930–1933, during which over forty talks took place annually. This corresponds with Hugo Steinhaus's opinion, according to whom "the Polish Mathematical Society progressed with uncommon vigor: sessions took place every week, and sometimes there were three or four talks in a given session" [B: Steinhaus, 2010; p. 121]. Ulam supported him in this view, providing some further details: "Sessions of the Polish Mathematical Society took place at the university almost every Saturday evening. Typically, three or four short communiqués were issued in the space of an hour, and then most of the attendees would depart for the café" [B: Ulam, 1969].

The number of talks delivered started to decline after 1934, though the number of sessions stayed more or less the same until the end, hovering more or less around thirteen to nineteen sessions annually. The decline turns out to be even more

[50]Many of these talks never appeared in print form.

apparent if one considers the fact that after 1934 the number of foreign guests participating in the sessions grew, as did the number of representatives of other Polish mathematical centers (and hence so too did the number of their talks). The increasing number of guests undoubtedly testifies to the attractiveness of the Lvov environment, by then enjoying a high international ranking. However, the reasons why fewer talks were given by representatives of the Lvov milieu seem to be complicated. One can argue that mathematical life had shifted towards the Scottish Café and that, to some extent, the *Scottish Book* had been assuming a substitutive role. Or one can point to the liquidation of the general studies department at PL and the departure of Kazimierz Kuratowski to Warsaw, or to the liquidation of the chair held by Stanisław Ruziewicz and his subsequent departure from UJK. Likewise, Stanisław Ulam and Mark Kac were making ever more frequent trips to the United States of America (where they later emigrated to), and Hugo Steinhaus had shifted his interests towards applied mathematics (he was head of the branch for the 1937/1938 term). Furthermore, there was a dramatic lack of available positions for young mathematicians. Finally, ever more ominous clouds were casting their shadow over public life (anti-Semitism, the threat of war).

The list of the ten most active speakers seems to be of interest: Stefan Banach, forty-three talks; Hugo Steinhaus, thirty-five; Juliusz Schauder, twenty-four; Stanisław Ulam, twenty-four; Władysław Nikliborc, twenty-two; Władysław Orlicz, twenty-one; Kazimierz Kuratowski, twenty; Stefan Kaczmarz, seventeen; Wacław Sierpiński, seventeen; Mark Kac, sixteen. Sierpiński was the only person on this list not to come from Lvov, while that is only partly the case with Kuratowski. Banach's activity drastically decreases from 1936 onwards, while that of Mazur becomes greater, his high position resting in large part on his joint involvement with the issuance of numerous communiqués, as is consistent with his reputation as a sharp commentator.

CHAPTER 12

Collaboration with Other Centers

The mathematics center at Lvov had a powerful way of drawing in others. The links between Lvov and Warsaw were particularly strong. Stanisław Ulam recalls them thus:

> From Warsaw Sierpiński would come occasionally, so would Mazurkiewicz, Knaster, and Tarski. In Lvov they would give short talks at the meetings of the mathematical society. In 1930 Mazurkiewicz spent a whole trimester lecturing in Lvov. Like Knaster in topology, he was a master at finding counterexamples in analysis, examples showing that a conjecture is not true. His counterexamples were sometimes very complicated, but always ingenious and elegant. Sierpiński with his steady stream of results in abstract set theory or in set-theoretical topology was always eager to listen to new problems—even minor ones—and to think about them seriously. Often he would send solutions, from Warsaw, to problems we had discussed in the café sessions. Similarly, Zygmunt came from Vilnius to visit us several times; he had always belonged to the Warsaw school from the start.
>
> We likewise received younger visitors from Warsaw. Borsuk, more my contemporary, came for a longer visit from Warsaw. We started collaborating from the beginning, and we published a number of papers. Leon Lichtenstein, a professor of Polish extraction based at Leipzig University, also visited for one trimester.
>
> Hurewicz, one of the most creative topologists, born in Poland, was studying abroad. He visited Lvov several times and gave talks at meetings of the mathematical society.
>
> (adapted from [B: Ulam, 1969])

Guest visitors from abroad also came, including Zermelo [Figure 15], H. Lebesgue, E. Borel, P. Montel, J. Leray, G. T. Whyburn. John von Neumann visited on two occasions and even wrote a problem into the *Scottish Book*. In 1938, UJK awarded Lebesgue an honorary doctorate. After the ceremony there was a party in his honor at the Scottish Café. Roughly fifteen people were present, which really does show just how few active mathematicians there were at the time (not only in Lvov). There is a little story about that party. Lebesgue was handed a menu printed in Polish but put it down after a few moments and declared: Merci,

FIGURE 15. Ernest Zermelo in Lvov (Zermelo gave a talk at the Lvov branch of the PTM on May 24, 1929). Seated (left to right) are Hugo Steinhaus, Ernest Zermelo, and Stefan Marcinkiewicz. Standing (left to right) are Kazimierz Kuratowski, Bronisław Knaster, Stefan Banach, Włodzimierz Stożek, Eustachy Żyliński, and Stanisław Ruziewicz.

> je ne mange que des choses bien definies[51]. To which Kac replied: Ne mangez jamais que des objets susceptible d'être definies par un nombre finis de mots[52].
>
> [B: Kac, 1978]

The personal ties brought results in the form of joint papers, investigations into certain types of equations, and saw the start of joint initiatives. One could hardly fail to notice that "Studia Mathematica" followed along the same lines as Warsaw's "Fundamenta Mathematicae" (a narrow scope of topics, papers would only be considered for publication if they were written in a language used at congresses, a promotion of the creative output of the community). However, the book series "Monografie Matematyczne" was a joint initiative. Of the ten volumes appearing in that series up to 1939, four were either prepared in Lvov or else had some strong connection with Lvov. W. Sierpiński and S. Mazurkiewicz worked in Lvov before World War I and felt quite at home there. W. Sierpiński shared particularly close bonds with S. Banach (H. Steinhaus called their meetings "banachalia"; see [B: Steinhaus, 1992; p. 94]) and S. Ruziewicz [B: Więsław, 2004].

[51]Thank you, but I only eat things which are well defined.

[52]Only eat things which can be defined in a finite number of words.

CHAPTER 13

In the Eyes of Others

Every mathematician is an individual, having not only specific intellectual gifts but also his or her own interests and viewpoints, both mathematical and nonmathematical. In such a close-knit community, a diversity of personalities creates a unique atmosphere and huge capacity to draw others in. Such a rich social fabric, pulsating with life and bountiful mathematical successes that were astonishing the world, was what the Lvov scene was all about. Stanisław Ulam sketched out an image of it all, based on his personal recollections after having spent thirty years in the United States:

> The kaleidoscope of mathematical types in Lvov was very diverse, not only with regard to interests and education, but also the type of intuition they had and their mathematical habits. The main driving force behind original research work at Lvov was work having a set-theoretical flavor: the fundamentals of set theory, set-theoretic topology, and then—under Banach and Steinhaus—functional analysis with applications to classical analysis. Schauder [...] focused on partial differential equations. His methods and his results have now become classical and remain some of the most powerful tools for achieving existence theorems. Banach, Mazur, and Schauder were the authors of what is now popularly referred to as the treatment of problems in analysis by exploiting geometric methods for function spaces.
>
> [B: Ulam, 1969]

Now that we have sketched out the ambience of mathematical life in Lvov, let us try to outline the individuals themselves, based on the recollection of people who knew them.

13.1. Stefan BANACH

> Stefan Banach, as I remember him, was a quiet boy, but not without a sense of humor, a good friend. He had a slightly secretive nature. He always wore a clean, proper uniform, as we all did; there were no signs of glumness or hunger, even though material circumstances forced him to give paid lessons to younger friends at school, as well as lessons in town; however, he helped his classmates for free.
>
> [B: Albiński, 1976]

FIGURE 16. Stefan Banach.

> [BANACH] was self-taught in mathematics and was already reading Tannery's book on real-valued functions while he was still at high school.[53]
>
> [B: Steinhaus, 1960]

> BANACH told me that after he and Wilkosz had passed their matura exam, they were convinced there was nothing much more to be done in mathematics because it had advanced so far. Therefore he later registered at Lvov Polytechnic, and not at the university.
>
> (A. Turowicz, quoted in [B: Pawlikowska-Brożek, 1990; p. 103])

> BANACH [...] was tall, had blond hair, blue eyes, with a somewhat heavy set build. His manner of speech [...] struck me by its directness, its power, and sometimes even by its excessive simplification; his facial expression was—as I asserted later—to

[53]Steinhaus goes on to write that "it is not known how Banach learned French." Actually it is: Jules Mien (1842–1905), a Frenchman by birth, was a friend of the family who brought him up in Cracow. He recognized the boy had talent and taught him his native language. Banach therefore spoke French quite well, though W. Sierpiński edited and translated some of his papers into French (see [B: Więsław, 2004]), while B. Knaster translated his monograph [A: Banach, 1932].

some extent consciously forced. [...] Typically his face showed he was in good spirits, together with a certain sceptical attitude. He tended to avoid direct confrontations in discussions; however, when he didn't agree with what someone was saying, it showed itself in the way he would pose questions. In mathematical discussions, into which he could be drawn very willingly if not eagerly, one immediately sensed the power of his mind. Whether it be in the university office or in the café, one could sit down with Banach for hours discussing mathematics problems. He drank coffee and smoked cigarettes almost incessantly. Those kinds of sessions with Banach, and more often with Banach and Mazur, ensured that the Lvov scene was unique in its kind.

[B: Ulam, 1969; pp. 52–53]

He [BANACH] did things his own way. He spent his free time away from lectures (of which he had particularly many) almost entirely in the café or at the restaurant. There, surrounded by friends and students, he could talk endlessly about mathematics, posing new questions, solving his own problems or those posed by other mathematicians. The café table became, beside the university departments, a place of inspiration for mathematical thought.

[B: Kuratowski, 1981; p. 88]

BANACH impressed people by his penetrating mind and the almost brutal force of his clear, but cold, intellect.

(Z. W. Birnbaum, quoted from [B: Woyczyński, 1997; p. 141])

BANACH liked to have fun. [...] In math lectures, at eight o'clock in the morning, Banach would sometimes come dressed in a tail coat, which looked pretty good on him, making him look a bit like a young diplomat [...]. The image of a reader wearing polished shoes while the flapping tails of his tail coat picked up chalk powder every other moment, with his brilliant white gorse shirt contrasting with the blackboard [...] was, if nothing else, an original one. Sometimes we noticed that the reader had neither combed his hair nor shaved himself. On one occasion he sat down at the table and rested his head on his hand. The room fell into total silence. [...] That is how we spent the time till the end of the lesson [...], not wanting to wake up the person in slumber.

(from the memoirs of S. Makowiecki,
quoted from [B: Hollanek, 1990; p. 84])

BANACH was above all a mathematician. He was not much interested in politics, though he had a sharp assessment on any matter brought to his attention. The natural world did not make any impression on him whatsoever; art, literature, and the theater were his secondary source of enjoyment, which would keep him occupied in his short, and very few, workbreaks—nonetheless, having a good drink in good company was something which mattered to him. Moreover, he could focus his mental effort on a single matter without being distracted by anything. He did not like to deceive himself, and he was fully aware that only a small percentage of people can understand mathematics. One day he told me: "You know what, my friend? The humanities are more important than mathematics at middle school—mathematics is too sharp an instrument; it's not for kids ...."

[B: Steinhaus, 1960; pp. 158–159]

Professor BANACH usually started giving his lectures in the middle of October, even though the academic year actually started on October 1. Banach started his holiday in September; he spent the summer months in the quiet of his office, and where it was draughty, like everywhere else in the "old 'versity", with its monasterial walls. Furthermore, I don't think Banach ever worked at home, at least only on extremely rare occasions. Rather, his workplace was either his office at the university or a special table at the Scottish Café. He spent his holidays in Skole for the summer, going by train along the Stryj-Ławoczne route. [...] Clarity of thought, a fluency with words, transparency, a calm manner of expressing himself, those were the main features of Banach's lectures, although not infrequently one did have the distinct impression that he never prepared his lectures, but he never got stuck because he quickly recovered his original train of thought, and the lecture continued.

[B: Szałajko, 1993; p. 47]

BANACH [...] was an extremely good lecturer; he never lost his way through the details and never covered the board with lots of complicated symbols. He was not overly concerned about verbal perfection; any kind of airs and graces was alien to him, and all his life he kept the manners of a Cracovian rogue in his behavior and speech. He found it very difficult to put down his thoughts on paper. He wrote his manuscripts on loose leaves of paper, ripped out from an exercise book; when he had to change part of the text, he would cut out the requisite places and glued in a new bit of paper, on which he wrote the new version. Without the help of friends and assistants, Banach's first works would never have reached the printers. He hardly ever wrote letters.

[B: Steinhaus, 1960; p. 155]

BANACH was similar to Hilbert in the sense that he attacked problems head on—after excluding the possibility of going along any side-paths, he concentrated all his effort on the one possible path that remained, leading straight to the goal—he believed

> that the logical analysis of a puzzle, much like how a chess player analyzes a difficult position, must lead either to a solution or to a refutation of the stated theorem.
>
> [B: Steinhaus, 1960; p. 154]

> [BANACH] used to say that mathematics rests on a specific kind of beauty and that it will never be possible to confine it within some rigid deductive system because every formal boundary confining it will be smashed, sooner or later, and new principles will be created. Of utmost importance to him was the theory of mathematics for its own sake, free of utilitarian considerations.
>
> [B: Steinhaus, 1960; p. 154]

The circumstances of war badly affected S. Banach's health. Jadwiga Hallaunbrenner recalls (quoted from [B: Kałuza, 1992; pp. 118–119]):

> Banach used to come to us regularly. My husband [...] got work in the Viehverband, a meat processing plant that dealt with the selling of oxen and the production of mediocre quality sausages. Every week he brought home a kilo of meat, namely some sort of inferior quality sausage or tripe. I used to roll down the window blinds in the kitchen, boiled up a pot of meat and beans, and in the early evening there would come to us Stefan Banach, Tadeusz Riedl, Kazimierz Brańczyk—a Polish literature specialist, a friend of my husband, and somebody else, whom I don't remember. Banach was exhausted, hungry, and emaciated, whereas before the war he was broad-shouldered and burly. When the men had finished eating they had discussions: about the situation on the war front, where the allies were likely to arrive, what was happening on the Eastern front, what was happening in Poland and in Lvov, what are the chances of survival, etc.

Banach's situation was dire not just because of the general hunger and the hopelessness of it, but also because of the type of work he was doing: during the years 1942–1943 his work required him to feed lice in Professor Weigl's institute, which produced antityphus vaccines for the Wehrmacht. The positive side of it was that, with all the endless checking of documents, his personal safety was guaranteed. The downside was that the work was physically exhausting. In these conditions, cancer developed in Banach's already weakened body, and it ultimately killed him one year after the Germans had been expelled from Lvov.

13.2. Hugo STEINHAUS

> STEINHAUS's approach [...] was marked by a deep appreciation of the historical development and continuity of ideas in mathematics. His book *What is and what is not mathematics*[54] influenced me greatly. Maybe it was because he was less interested

[54] The book was published in 1923 and was never revised.

in highly abstract mathematical fields and less feeling for them that he tried to guide our thoughts to more practical applications [...]. He had a fondness, and a talent, for finding problems [...] grasping hold of something [...] which could be formulated in a mathematical way. He [also] had a particular gift for language. He sometimes, rather pedantically, required us to express ourselves precisely when writing or talking about mathematics or mathematical branches of the natural sciences.

[B: Ulam, 1969; p. 54]

FIGURE 17. Hugo Steinhaus.

Professor STEINHAUS had, what I may somewhat pretentiously call, something in common with Socrates' method. Namely, he liked to lead people into a deeper understanding by posing questions. And he posed questions which pinned them down, so they didn't know what further to say. I believe he recalled in one of his stories from his Göttingen days when he was talking with his landlady and asked her why people pay taxes. She replied, "Well, one pays taxes to maintain the army." "But what is the army for?" "Well, to fight the French." "But why?" She had no answer to that. He used that approach very often.

Professor STEINHAUS liked to instruct younger people to maintain a certain orderliness and precision in their work. I shall relate what I heard from Professor Mazur, who while still a student had written his first mathematical paper and presented it to Professor Steinhaus. The latter was supposed to introduce it at a meeting of the Lvov Scientific Society. The meeting was supposed to take place in the evening. Just a few hours before it was due to begin, Professor Steinhaus summoned Mazur and said, "Mr. Mazur, how can I talk about your paper when you haven't given it to me yet?" Mazur was dumbfounded. "I don't know what you mean; I handed it in to you." "No, you did not. You handed me four sheets of blank paper." The latter was horrified, and then Professor Steinhaus continued, "Here you are; see for yourself." What had happened was this: before the war one could get a certain type of really cheap quality paper. It was yellowish in color. Mazur's inkpot had run dry, but he didn't want to go to the shop for more, so he just added water. Professor Steinhaus took Mazur's work and says, "Well, indeed, something is actually written here. However, if you want to do academic work, I would advise you to stick with white paper and black ink."

[B: Turowicz, 1973; pp. 87–88]

His [STEINHAUS's] lectures were always meticulously prepared, unusually precise, and delivered with a beautiful use of language. [...] He did not give advance notice of the start date of his lectures, though everybody knew it was always on October 1, unless it was a Sunday.

[B: Szałajko, 1993; p. 47]

Let us return to Professor STEINHAUS's lecture. Perched high on a podium beside a tiny (roughly 1.5 m x 1.5 m) blackboard, his figure dominated over the room. The lecture was always planned down to the smallest detail and written out. While lecturing he held in his hand an exercise book with his notes. The dimensions of the blackboard meant he was perpetually erasing it. His lectures were very energetic, delivered with fluency, while adopting a certain pose. Despite the meticulous preparations, the lecture was too difficult for most students. In that respect he was the complete opposite of Banach, who had carefully prepared lectures, yet he delivered them in such a manner that I don't think there was anyone who didn't understand everything he was saying. Professor Steinhaus often used to forget assumptions, or else he used to lose his way when forced to use equations he had already erased earlier. However, the most important moments were when the professor lost his way. Then we learned how to do proper mathematics. That was when the true professor in him revealed itself.

As a fresher student I mainly saw him during lectures, in his office, or in the corridor. He usually strode around haughtily, deep in thought, very stylish, a seemingly aloof and unapproachable figure. However, the moment he lost his way in a lecture, that pose of his completely vanished, and there stood before us a senior colleague, engaged in animated discussion with us, and only too pleased if we gave him a good idea. At that moment there was no distance between us, he showed us what he was really like, he was the person I would get to know a short while later. Those were the most interesting moments in a lecture.

[B: Stark, 1973; p. 78]

13.3. Stanisław MAZUR

MAZUR was the person who [...] taught me to control my inherent optimism—how not to rush too quickly to conclusions on the basis of an outline proof, without checking through the exact details. [...] Mazur's great strength was in what he himself called making observations and "comments." They typically captured, in an uncommonly concise and precise manner, the specific properties that stem from a given notion. Once they were noted, they weren't that difficult to check, but in general they were "na terenie" [on terrain] which was hard for most mathematicians to notice much. Very often his observations were the key insight to proving a theorem or finding a counterexample.

Let me talk about some of those insightful thoughts which later became the subject of many mathematical papers. During a café discussion, he gave the first example of an infinite mathematical game. Let me remind you that the year was 1929, or 1930, when Mazur touched on the subject of automatons capable of answering questions about themselves, so long as they have a certain amount of available material in their immediate vicinity. We discussed this in quite abstract ways, and some of our thoughts, which were never written down, were actually precursors to later work.

[B: Ulam, 1969]

FIGURE 18. Stanisław Mazur.

Stanisław MAZUR's lectures were precise, without extraneous words, and delivered in a tranquil tone.

[B: Szałajko, 1994, p. 258]

13.4. Kazimierz KURATOWSKI

KURATOWSKI and Steinhaus, each in a different way, represented elegance, rigor, and intelligence in mathematics. Kuratowski was really a representative of the Warsaw school which flourished almost explosively after 1920. He came to Lvov in 1927, preceded by a reputation for his work in pure set theory and axiomatic topology of general spaces. As editor of Fundamenta Mathematicae he organized and gave direction to much of the research in his famous journal. His mathematics was characterized by what I would call Latin clarity.

[B: Ulam, 1969]

FIGURE 19. Kazimierz Kuratowski.

13.5. Stanisław ULAM

Stan took to café mathematics like a fish to water. In the casual ambience of the Scottish Café Stan blossomed into one of the most promising mathematicians of his generation. He also began to display the contradictory traits in behavior that after his operation[55] were to become dominant: deep intuition and impatience with detail, playful inventiveness and dislike of prolonged work. He began to view mathematics as a game, one that a well-bred gentleman should not take too seriously. His insights have opened whole new areas of mathematics, all of them still actively cultivated today, but he himself could not bear to give his discoveries more than a passing interest, and at times he would make merciless fun of those who did take them too seriously.

Had he been able to remain in Poland and survive the war, as Steinhaus, Kuratowski, and others did, he would have gone on to become one of the leading international figures of pure mathematics, at least on a par with Banach. But after he bade farewell to his friends at the Scottish Café, something died forever within him.

[55]In 1945 Ulam survived major brain surgery, but it caused a severe personality change in him: in particular, he lost the capacity for sustained concentration.

FIGURE 20. Stanisław Ulam.

> Stan arrived in the United States ill equipped for the rigor of puritan society. The big open spaces of America, the demands for aloneness and self-reliance made him feel estranged. He wished to belong [...] but he never came to feel fully at home [...]. He missed the lively street life of European cities, the culture, the rambling conversations [...] and viewed with alarm the decay of that art ....
>
> (all from [B: Rota, 1990])

13.6. Juliusz SCHAUDER and Zbigniew ŁOMNICKI

The number of professors both at the institute and the university was extremely limited and their salaries were very low[56]. People

[56]The salary of an ordinary professor or department head (group IV salary level) was reasonably acceptable, amounting to 1,000 zloties, at group V salary level it was 700 zloties (professor extraordinary), and at group VI level, it was 450 zloties [Statistical Yearbook, 1939]. Ulam's note undoubtedly refers to readers, who were not on a stable salary scheme, as well as to assistants and adjuncts.

FIGURE 21. Juliusz Schauder.

like Schauder had to teach in high school to supplement a meagre income as lecturer or assistant. Zbigniew Łomnicki worked as an expert in probability theory in a government institute of statistics and insurance. Despite all that, most mathematicians found the time to frequent the cafés.

(adapted from [B: Ulam, 1969])

13.7. Włodzimierz STOŻEK

STOŻEK, who was a professor and chairman of the Department of General Studies at Lvov Polytechnic, used to come to the café every day, like many others. Stożek, short, rotund, completely bald, and always in good humor, used to play chess with Nikliborc, who was a reader and a senior assistant at the same time. They used to spend the afternoons playing chess in the café, surrounded by other mathematicians who were afficionados.

(adapted from [B: Ulam, 1969; p. 55])

> The lecture [given by Włodzimierz Stożek] commissioned for the natural scientists at the university [...] was very easy to understand; indeed it was at just the right level for natural scientists. Moreover, it was conducted in a very easy going and humorous way.
>
> [B: Szałajko, 1994; p. 254]

> STOŻEK was quite an easy going person, generally cheerful, and with a good sense of humor.
>
> [B: Szałajko, 1994; p. 263]

13.8. Herman AUERBACH

> AUERBACH, who was also a good chess player, often came there. He had been a senior assistant, later promoted to a readership. Outwardly timid, he was often capable of very caustic humor. His mathematics was elegant, and indeed it was from him that I learned some rather classical mathematical fields.
>
> [B: Ulam, 1969; especially page 42]

> AUERBACH, a reader, was a very kind person, very modest, helpful, and much liked by students. By virtue of his huge wealth of knowledge, students called him the walking encyclopedia, because he really was like that. Stefan Banach thought so too, and sometimes he sent students to go and see Auerbach for a consultation and to find out from him what the relevant literature was.
>
> [B: Szałajko, 1994; p. 252]

13.9. Others

KACZMARZ, tall and thin, a friend of Nikliborc, turned up [at the café] from time to time. ORLICZ, who likewise was an assistant at the university as well as a friend of Mazur, tended to come less frequently.

RUZIEWICZ, a professor at the university, came to the café often, just like artistocratically looking Antoni ŁOMNICKI. The latter was a professor at the polytechnic, a probability theorist, an authority on cartography, and the author of a good textbook. His nephew, Zbigniew Łomnicki, later became my good friend and mathematical collaborator. My teacher, KURATOWSKI, and STEINHAUS too, only turned up at those café sessions on the odd occasion. They usually frequented a rather upmarket coffeehouse located not far from our café.

The most active and most productive of the younger lot at that time were myself and SCHREIER, a university student. The series of papers we published back then testifies to our near-everyday mathematical collaboration. [...]

Mark KAC, who was about four or five years younger than Schreier, was studying at the university under STEINHAUS. Already as a freshman he evinced singular talent and obtained important results, initially working together with STEINHAUS on probability theory and the theory of Fourier series. [...]

I remember some of the professors of other subjects—chemistry, physics, and so forth—who came and sat down with us either at the café or at the coffeehouse. MAŁACHOWSKI, a really humorous professor of chemistry, sometimes used to play chess with AUERBACH and STOŻEK [...]. AJDUKIEWICZ, the philosopher, came regularly. ŻYLIŃSKI, who resembled a cavalry officer more than a university professor, appeared from time to time on short visits [...]. In fact, just about everyone who worked as an academic either at the university or the polytechnic turned up at the café at one time or other.

Note: This section is adapted from Ulam's text (English version), [B: Ulam, 1969].

Part III. THE GOLDEN AGE
Achievements

CHAPTER 14

Stefan Banach's Doctoral Thesis and Priority Claims

If Stefan Banach was Hugo Steinhaus's "greatest mathematical discovery" (see Chapter 6), then his meeting with Steinhaus was equally influential, if not decisive, for Banach himself. He began studying mathematics in a regular, more disciplined way, whereas before he had been more of a keen enthusiast. Steinhaus set for him a problem in the year they first met, namely 1916 (the problem: is there an integrable function which does not converge on average as a trigonometric series?). Banach sketched out a solution in just a couple of days, showing that such a function cannot exist. Steinhaus filled in a few gaps in the proof, and that became their first joint paper [A: Banach & Steinhaus, 1918] and Banach's first ever paper. Strictly speaking, Banach was still a student at the time it was printed. In the next few years Banach published several more papers on trigonometric series and functional equations. We shall mention two of them here. The paper [A: Banach, 1920] concerns a well-known problem of Cauchy about finding a real-valued function which satisfies the equation $f(x+y) = f(x) + f(y)$ for all x, y. It is an example of a *functional equation* (because the function f itself is unknown). In the case here, one has to find an additive function, which is the name given to functions satisfying the above equation. Cauchy knew that if an additive function f is continuous, then it must have the form $f(x) = cx$, where c is a constant. Banach showed the same conclusion is true for additive functions that are measurable in the Lebesgue sense (a much weaker assumption than continuity). In [A: Banach & Ruziewicz, 1922] the authors found an elementary way of obtaining solutions f, ϕ to Maxwell's functional equation $f(u)f(v)f(w) = \phi(u^2 + v^2 + w^2)$.

Thus S. Banach's first papers were concerned with well-understood trigonometric series and functional equations. They paved the way for later papers about spaces consisting of functions, about measure and measurable functions (in the Lebesgue sense), about additive functions, and about the axiom of choice. Already one can find in them characteristic features of S. Banach's later creativity: familiarity with the latest literature and the intellectual audacity to apply new methods (such as the axiom of choice), an interest in measure and function spaces, consumate skill in delivering proofs, elegant reasoning, and a natural ease when working with others.

Banach's talent quickly unfolded. The first flashes of his genius were seen in his doctoral thesis of 1920 (A. Łomnicki was his thesis adviser), published two years later [A: Banach, 1922a][57]. Its greatest achievement was in the definition of a

[57]Banach cared much more about getting results than getting things published, and this might explain the delay in publication.

general abstract space which included several of the function spaces already known at the time (see Chapter 5).

How should one look for such a generalization? Banach was clearly guided by geometry. He looked for it by seeking a generalization of the real line and, even more generally, of Euclidean space itself. That is to say, he sought to define a space which would preserve the most important properties of Euclidean space, thereby allowing him to apply known geometric methods to the new space and thereby extend classical analysis to it. His genius lay in recognizing that known function spaces enjoyed such properties of Euclidean spaces as linearity, the existence of a metric, and completeness. Linearity is an algebraic property. The possibility of measuring distances (metricality) can be thought of as an analytical property because it provides a way of estimating how close one point is to another: this then leads to a most fundamental notion in analysis, namely the limit point. Completeness is a topological notion: it is a statement, in a certain sense, that a metric space does not have holes in it. These three properties had to be linked to each other in some way (and simply perceiving all this required the spark of true genius). This is how algebra, analysis, and topology became fused together in Banach's conceptual thinking, but the unification he sought appeared to run in a direction counter to that indicated by geometry. Very important in geometry is the notion of angle, and consequently that of orthogonality (perpendicularity). It can be expressed via the concept of a scalar product. Banach's greatness can be seen in the way he was not tempted to require his general space to be endowed with any kind of scalar product whatsoever: this of course impoverished its geometry but it also guaranteed greater generality. He decided merely to stick with the more modest condition of metricality. Time would show that it was the right decision.

To acquaint ourselves with Banach's definition of a more general space, which we shall do shortly, it is good to have some illustrative examples before us. Consider any of the following: a set of numbers (for example the space $\mathbb{R}$ of real numbers), a set of points (for example a set of points in Euclidean space $\mathbb{R}^n$), a set of sequences (for example the set ℓ^2 of sequences x_n such that $\sum x_n^2 < \infty$), a set of functions (for example the set $C[0,1]$ of continuous, real-valued functions defined on $[0,1]$) etc. The definition, as given below, of a type (B) space is a generalization of all these spaces, as well as many others, and it retains all their basic properties.

Banach's definition was axiomatic. He broke his axiomatics up into three groups corresponding to the three properties described earlier, namely linearity, metricality, and completeness. The first group was the largest. It stipulated that his space X should be a vector space over $\mathbb{R}$. That is to say, for elements of X, referred to as vectors, addition must be defined, with respect to which X is an abelian group, as well as multiplication of vectors by real numbers (which, in this case, are sometimes referred to as scalars), satisfying the conditions for associativity and distributivity. Specifically, for any given elements $x, y \in X$, an addition rule $x + y \in X$ is defined, which is both associative and commutative:

$$x + (y + z) = (x + y) + z, \quad x + y = y + x \quad \text{for all } x, y, z \in X.$$

X contains a neutral element 0 such that $x + 0 = 0$ for all $x \in X$. Furthermore, for every element $x \in X$, there exists an *inverse* element $-x \in X$ such that $x+(-x) = 0$.

The second group of axioms was characterized by the norm function of vectors in X. It is denoted $||x||$, for $x \in X$. It is a real-valued function and it satisfies the following conditions:

a) $||x|| \geq 0$,
b) $||x|| = 0$ if and only if $x = 0$,
c) $||ax|| = |a| \cdot ||x||$,
d) $||x + y|| \leq ||x|| + ||y||$.

The underlying reasoning is geometric in nature: the norm of a vector stands for the intuitive notion of the length of a vector, while the last of the conditions stands for the so-called triangle inequality, which states that the sum of the lengths of the two shortest sides of a given triangle is greater than or equal to the length of the longest side. The existence of a norm makes it possible to map X into a metric space: the distance ρ between two elements x, y of the space X is defined as the norm of their difference, namely $\rho(x, y) = ||x - y||$. A simple check confirms that any function ρ satisfying these conditions has the properties of a metric, that is to say, if, for each $x, y, z \in X$, the following conditions are satisfied:

$\rho(x, y) \geq 0$,
$\rho(x, y) = 0$ if and only if $x = y$,
$\rho(x, y) = \rho(y, x)$,
$\rho(x, y) \leq \rho(x, z) + \rho(z, y)$.

The third group consists of but a single axiom, about completeness. It stipulates that if $\{x_n\}$ is a Cauchy sequence with respect to the norm, meaning that $\lim_{n,m \to \infty} ||x_n - x_m|| = 0$, then there exists an element $x \in X$ such that

$$\lim_{n \to \infty} ||x_n - x|| = 0.$$

Stefan Banach called any abstract space satisfying the three groups of axioms described above a *type* (B) *space*, but a few years later it was being called a Banach space[58] and is called such today. In a nutshell, a Banach space is a vector space defined over the real numbers $\mathbb{R}$, equipped with a norm and, furthermore, complete with respect to its norm. Even more succinctly, it a normed and complete vector space. Although a Banach space does not require the geometric notion of a scalar product, and hence no orthogonality notion, Banach's definition turned out to be the right one: apart from the examples given earlier, the spaces which bear his name include the space L^2 of square-integrable functions on the unit interval $[0, 1]$, the space L^p (for which no scalar product exists, in general), the space of measurable functions, and many others besides. All of them can be treated within a common framework, yet in quite broad generality, leading to important theorems. It truly was the beginning of a new branch of mathematics. In that same year of 1922, Paul Lévy (1886–1971) named it *functional analysis*[59]. It should also be noted, however, that although it was only natural for Banach to consider vector spaces over the real numbers, that was a major limitation, and many of the results obtained by the Lvov school were later extended to include vector spaces over the complex numbers $\mathbb{C}$.

---

[58]The name is attributable to Hugo Steinhaus who, quoting M. Fréchet, was the first person to call a type (B) space a *Banach space* [A: Steinhaus, 1929]. The name quickly stuck and remains to this day.

[59]The name comes from the title of the book [C: Lévy, 1922].

Stanisław Mazur described it thus [B: Mazur, 1961; pp. 249–250]:

> Functional analysis arose, just as every new academic discipline arises, as the end phase of a long historical process. The list of people whose research contributed to the establishment of functional analysis is long; it includes names such as Vito Volterra, David Hilbert, Jacques Hadamard, Maurice Fréchet, and Fryderyk Riesz. But 1922, the year when Stefan Banach presented his doctoral thesis entitled *Sur les opérations dans les ensembles abstraits et leurs applications aux équations intégrales* in the Polish journal Fundamenta Mathematicae, is a breakthrough year in the history of twentieth-century mathematics because that thesis, running several dozen pages, finally laid down the foundations of functional analysis [...].
>
> Functional analysis replaced the basic notion of number in mathematical analysis by a more general notion, nowadays referred to in thousands of mathematical theses as a "point in Banach space". Having achieved this generalized mathematical analysis, called functional analysis, one could treat radically different problems in mathematical analysis in a simple and uniform way and use it to find solutions to problems which mathematicians had previously struggled with unsuccessfully. It provided academics working in the natural sciences, particularly physicists, with more powerful mathematical tools.

Banach's thesis was not confined merely to the definition of a type (B) space. It went much further. It was still necessary to show that the definition fulfilled certain expectations and defined an interesting mathematical object in its own right. And Banach did exactly that. One could even go so far as to say the principal subject of his doctoral thesis was not so much the definition itself of an abstract space, but rather the definition of linear operators $u : E \to F$, where E, F are arbitrary normed vector spaces having the completeness property with respect to their norms (in other words, they are Banach spaces), together with the sequential limits of such operators. In developing the theory of such operators, he often stipulated that they be additive (alternatively, linear; in the Euclidean case the only linear operators are translations, rotations, and reflections, or their combinations) and he used the language of geometry and extended it greatly for the purposes of his theory.

In particular, having defined his notion of a space and the operators between the different kinds of spaces and after proving some theorems about them, Banach then made a distinction between continuous operators (an operator F is said to be continuous at x_0 if $F(x_n) \to F(x_0)$ whenever $x_n \to x_0$), uniformly continuous operators (where the convergence is uniform for all points x_0 in the space), and additive operators ($F(x+y) = F(x)+F(y)$ for all x, y). He also defined convergence of operators ($F_n(x) \to F(x)$ for all x).

He proved many theorems in the later part of his thesis. Here we shall mention only two of them. The first one, which had much importance for the emerging theory, states that the limit of a pointwise convergent sequence of linear, continuous operators is itself a continuous linear operator. The second one, being a generalization of the frequently used method of successive approximations, provides a practical way of finding solutions to various types of equations. Known as the contraction principle or as Banach's fixed point theorem, it has many applications.

Banach's Fixed Point Theorem. *Let F be a continuous operator mapping a Banach space X to itself. If there exists a number M satisfying $0 < M < 1$, such that $||F(x') - F(x'')|| \leq M||x' - x''||$ for all $x', x'' \in X$, then there exists a unique element x such that $F(x) = x$.*

The theorem is consistent with intuition: if an operator F maps a Banach space to itself and is "a contraction mapping", namely if it makes points come closer together (expressed by the strict inequality $||F(x') - F(x'')|| < ||x' - x''||$), then a certain (special) point never gets moved (whence the claim $F(x) = x$).

The value of S. Banach's thesis is that it provided clear notions which could be used (which was not yet obvious at the time) as the basis of a rich theory, one which synthesized all the previously established results on function spaces, and one which promised lots of applications. In any event, the notion of a Banach space prompted various questions. Some of them were posed and solved by Banach himself, while other Lvov mathematicians followed in his footsteps. This is how classical Banach theory was born, and with it, so too was the Lvov School of Mathematics.

Before the theory began to develop, however, there arose priority issues. During 1920–1922 concepts similar to those of Banach were formulated by N. Wiener and H. Hahn. Wiener, who at that time was working on logic, described his concept in terms of a complicated and highly abstract logic system, but he did not explain the motivation behind it and he did not provide any examples [C: Wiener, 1921]. Hahn, on the other hand, was guided by the ideas of E. Helly about solving infinitely many systems of linear equations in infinitely many variables, but it was all expressed in the language of sequence spaces [C: Hahn, 1922]. So the ideas that motivated each of them were radically different. One might say, possibly with a little oversimplification, that Banach was guided by geometry and his definitions were clear, simple, and transparent—much like a crystal structure. Wiener was guided by logic, but his definitions were unclear. Hahn was guided by algebra, but one had to be fluent in the language of systems of linear equations in order to extract the wider meaning of what he was doing. Nobody at all was interested in Wiener's paper (it is not even cited in historical works about functional analysis; see [B: Dieudonné, 1981]), and Wiener himself quickly transferred his interests to other areas, even writing about this episode later [B: Wiener, 1958; p. 60] (the paragraphs below are taken verbatim from the original text):

> However, I have a full set of axioms for vector spaces. Fréchet liked it, but did not seem particularly struck with the result. But then, a few weeks later, he became quite excited when he saw an article published by Stefan Banach in a Polish mathematical journal which contained results practically identical to those I had given, neither more nor less general. Banach's conception of his ideas and his publication of them were both a few months earlier than my own. There had, however, been no chance of communication between us, and the degree of originality of the two papers was identical.

> Thus the two pieces of work, Banach's and my own, came for a time to be known as the theory of Banach-Wiener spaces. For thirty-four years it has remained a popular direction of work. Although many papers have been written on it, only now is it beginning to develop its full effectiveness as a scientific method.
>
> For a short while I kept publishing a paper or two on this topic, but I gradually left the field. At present these spaces are quite justly named after Banach alone. There were several motives which led me to abandon this brain child, of whom I was at least one of the parents. The first was that I did not like to be hurried or watch the literature day by day in order to be sure that neither Banach nor one of his Polish followers had published some important result before me. All mathematical work is done under sufficient pressure, and its increase by such a fortuitous competitive element is intolerable for me.

Thus, N. Wiener himself acknowledged that Banach had priority [C: Wiener, 1923], and he went on to pursue other things (Brownian motion, in particular).

The situation was different with H. Hahn. Although Hahn and Banach were not in contact with each other and worked with different structural notions, they did share similar intuitions, and their work overlapped each other's (see Chapter 18). Nonetheless, Banach's contributions were the greater of the two because his notions were usually simpler and his theorems more elegant [B: Duda, 2005a]. By the time of H. Hahn's untimely death in 1934, Banach was already the unquestioned, supreme authority on the spaces which bear his name.

CHAPTER 15

Probability Theory

Probability theory is a relatively young branch of mathematics. The beginning of *probability calculus*, as it was better known for a long time, is considered to date back to 1654, the year when B. Pascal and P. Fermat exchanged letters about their work on the laws governing dice throws and the chances of winning dice games. In 1657 Ch. Huygens published his treatise *De rationiis in ludo aleae* on games of chance (gambling), while in 1713 Jacob Bernouilli's famous *Ars conjectandi* appeared, containing the first definition of probability: "probability is a degree of confidence, and appertains to the degree of certainty to which a part belongs to the whole" (quoting from [B: Urbanik, 1973]). P.-S. Laplace was the first person to formulate the so-called classical definition of probability, which assumes there are only a finite number of possible outcomes, each outcome being equally likely to occur as any other, and the probability that a given event occurs is expressed as the ratio between the number of outcomes in which the given event may occur and the total number of possible outcomes. His 1812 treatise entitled *Théorie analytique des probabilités* marked the beginning of probability calculus as a distinct branch of mathematics, concerned with quantities that obey the laws of chance. The contradiction between the "laws of chance" and the intuitive understanding of chance, as a random and unpredictable event, remained bothersome, and in the nineteenth century the probability problem was taken up by the finest mathematical minds of the age: P.-S. Laplace, A. de Moivre, C. F. Gauss, S. D. Poisson, P. Chebychev, J. Bertrand, H. Poincaré, A. Markov, and others. They enriched and made considerable progress with classical probability calculus, but they did nothing to address its underlying foundations. We should add that Jan Śniadecki (1756–1830) favored *calculus probabilitatis* in Poland and called it the "calculus of hit and miss" ["rachunek chybi-trafi"]; later (in Vilnius) he called it the "calculus of chance events" ["rachunek losów"].

It was also in the nineteenth century when the physicists R. Clausius, J. C. Maxwell, and L. Boltzmann showed how notions such as the kinetic theory of gases, like pressure, temperature, or entropy, can be interpreted as the statistical consequences of collisions between molecules experiencing continuous motion. They thereby reduced the laws of thermodynamics to theorems in probability calculus. As a consequence, there appeared new, and different, applications and questions to consider in probability calculus, ones which were distinct from games of chance (and, more generally, distinct from the classical case). In its day, the 1912 treatise of J. Gibbs, entitled *Elementary Principles in Statistical Mechanics*, was massively influential for the development of theoretical physics based on probability calculus. However, the nature of probability itself, the actual foundations which could justify the calculations, still remained unclear.

In his problem VI from 1900 Hilbert asked [B: Więsław, 1997] for the axiomatization of those parts of physics in which mathematics has played an essential role, in the first place of mechanics and of probability calculus (the latter was thus for him a part of physics). There followed some, rather unsuccessful, attempts by G. Bohlmann (1908) and S. Bernstein (1917), frequency conception of R. von Mises (1919), and E. Borel's paper (1909) in which he came up with an idea of a proximity between the probability calculus and the theory of measure. All that, however, has been unsatisfactory and the problem of basis for the probabilty calculus remained open.

H. Steinhaus became interested in measure theory quite early on. Already in 1920 he had produced a paper in which he observed that if one takes a set of positive measure and considers the set of all possible distances between its points, then this latter set contains an interval of the form $[0, a)$, for some real number $a > 0$ [A: Steinhaus, 1920; Th. VIII]. As is sometimes the case in mathematics, the actual proof was not hard. The real difficulty was in formulating the problem in the first place, a problem bordering on metric topology and measure theory. Later on, that paper would be cited often. Among other things, S. Ruziewicz showed that it is possible to insert into every subset, of positive measure, of the real line an arbitrary, finite linear configuration of points correspondingly diminished [A: Ruziewicz, 1925].

In 1923 two important papers were included in the same volume of "Fundamenta Mathematicae". The authors, both Polish mathematicians, were A. Łomnicki (paper submitted in 1920) and H. Steinhaus (paper submitted in 1922)[60]. The first paper [A: Łomnicki, 1923] tried to base probability calculus on measure theory. The author accordingly defined the notion of an elementary event and a random event. He also stated a definition of probability itself as a measure defined on random events. He supported his conceptual viewpoint with examples.

Łomnicki's paper went virtually unnoticed. One possible reason might have been that the author did not specify notions such as random variable or expected value. Furthermore, he completely omitted the important notion of independence, namely a notion which makes a different weighting between probability calculus and measure theory. Nevertheless, it constituted important progress towards establishing solid mathematical foundations for probability calculus.

Having long been interested in power series, H. Steinhaus had asked himself, while still in Jasło, what the probability was that a power series would converge if its coefficients were random variables [B: Steinhaus, 2010; p. 104]. Those thoughts bore fruit in an important paper [A: Steinhaus, 1923a]. Its importance rested on the way he presented a full mathematical framework for the important case of a nonclassical probability system, namely the game of heads or tails. More precisely, he formalized the possible outcomes of a game of heads or tails as sequences of zeros and ones, thereby representing them (in binary form) as real numbers lying in the unit interval $[0, 1]$. Random events were Lebesgue measurable sets, and their probabilities were defined as their Lebesgue measures. His approach had an axiomatic character; his axioms for nonterminating coin-tossing games were based on Sierpiński's axiomatics for Lebesgue measure. As a consequence, the isomorphism between that game and Lebesgue measure on the unit interval $[0, 1]$

[60]Several articles discovered the importance of these papers, including [B: Urbanik, 1973], [B: Girlich, 2005], [B: Shafer & Vovk, 2006].

became apparent. H. Steinhaus demonstrated the power of this formulation by solving one of his own problems[61] regarding the convergence of infinite series in which the constituent terms are random variables. Specifically, the probability of convergence of a series of real numbers $\sum \pm c_n$, assuming the signs occur randomly with an equal probability of $\frac{1}{2}$, is equal to the Lebesgue measure of those numbers t such that $\sum c_n f_n(t)$ is convergent, where $f_n(t)$ is a function defined on the unit interval $[0,1]$, and assumes the values 1, -1 on intervals defined by points $k/2^n$, where $k = 0, 1, \ldots$. Even more formally, the functions $f_n(t) = \operatorname{sign}(\sin 2n\pi t)$ are known as the Rademacher functions. Rademacher functions are an example of a sequence of independent functions. H. Steinhaus placed much importance on the role on independence, so he set up the "Polish school of independent functions"; see [B: Holgate, 1997].

In contemporary parlance, H. Steinhaus had constructed a probability space for the game of heads or tails:

$$([0,1], L, \lambda)$$

where $[0,1]$ is the unit interval, L is the family of all Lebesgue measurable subsets lying on this interval, and λ is the Lebesgue measure. It was the first mathematical axiomatization in which random events were expressed in terms of a well-defined probability space, though admittedly it was restricted to a very specific probabilistic scenario—the game of heads or tails.

H. Steinhaus's result was generalized for sequences of independent random variables [C: Khintchine & Kolmogorov, 1925], but it still did not constitute the desired axiomatization of probability itself. That definitive step was made ten years after H. Steinhaus, by A. Kolmogorov [C: Kolmogorov, 1933]. It depends on an extension of Steinhaus's ideas: Kolmogorov replaced the head-or-tails sequence with arbitrary random variables, and he replaced the Lebesgue measure with a normed measure. Ever since then, the triple

$$(\Omega, \mathcal{F}, \mathbb{P})$$

where Ω is the space of elementary events, $\mathcal{F}$ is a σ-field of subsets of Ω (random events), and $\mathbb{P}$ is a normed measure defined on the σ-field $\mathcal{F}$, became the generally accepted mathematical notion defining random systems. Another important feature of A. Kolmogorov's results is that it afforded a measure-set-theoretic interpretation of stochastic independence. As a consequence Hilbert's VI problem, which partly concerned probability, was solved.

> In the language of the sports commentator, one can say Kolmogorov got to the final, but the two Polish mathematicians H. Steinhaus and A. Łomnicki got to the semifinals. Their achievement is all the greater in that they, together with Wacław Sierpiński and Andrzej Mostowski (whose work had led to the solution of Hilbert's first problem—the Cantorian continuum hypothesis) were the only Polish scholars whose contribution to the solution of Hilbert's problems was of world class stature.
>
> [B: Urbanik, 1973; p. 45]

In any case, starting from [A: Steinhaus, 1923a], H. Steinhaus and his students were already thinking about probability theory in a modern way, using terms from measure theory. That was pioneering progress for the time. Later W. Feller made

[61] A. Khintchine won the 1941 Stalin Prize for solving a different problem from that work.

it common to a wider readership, but at the time he knew nothing of the work done by A. Łomnicki and H. Steinhaus. He only found out about Steinhaus's work from the author himself in 1937 at a conference in Geneva [B: Steinhaus, 2002; p. 161]. Steinhaus himself applied his approach to Taylor series having random terms, and R. E. A. C. Paley and A. Zygmund followed on from his example with a paper in "Studia Mathematica" 2 (1930) on random Fourier series.

To be explicit, what Steinhaus did [A: Steinhaus, 1930a], [Steinhaus, 1930c] was to look at sequences $\Theta_1, \Theta_2, \ldots$ of selections $\Theta_i \in [0,1]$. He constructed an axiomatic framework analogous to the one he had considered earlier for $\Theta_i = 0$ or 1 [A: Steinhaus, 1923a]. He showed the existence of an isomorphism between the set of these generalized sequences and the Lebesgue measure on the unit interval $[0,1]$. This time he exploited a type of Peano mapping from the product space $[0,1]^\infty$ to the interval $[0,1]$. He applied this isomorphism to prove some limit theorems.

Steinhaus's axioms were measure-theoretic in nature, but he suggested they might be applicable to arbitrary sequences of random variables. S. Ulam made that step and delivered a communication at the Zurich Congress stating that he, working jointly with Z. Łomnicki, had constructed a product measure on abstract spaces [A: Ulam, 1932]. They cited Kolmogorov's book [C: Kolmogorov, 1933] in their joint paper [A: Łomnicki & Ulam, 1934], while A. Kolmogorov cited their paper in his talk at the All Union Congress of Mathematicians in Leningrad (June 1934); see [B: Shafer & Vovk, 2006; p. 86]. The joint paper of Z. Łomnicki and S. Ulam was an important one but was not as influential as it could have been. One reason was because the authors gave a fallacious proof of their stated theorem on product measures (but it should be noted that others made the same mistake as well).

> Steinhaus's mathematical legacy, as well as those of Paley and Zygmund, was quickly picked up on in France by Salem, Billard, Kahane, Pisier, Ledoux, and Talagrand. Steinhaus would have been amazed and overjoyed if he could have lived to see the way the two theories became interconnected, having been developed from just a few basic ideas, materializing in print in the form of a massive book by Ledoux and Talagrand (published in 1991, 500 pages, 600 references): *Probability and Banach Spaces*[62]. On the other hand, M. Kac considerably enlarged Steinhaus's legacy in the most diverse directions of probability theory and its applications.
>
> [B: Kahane, 1995; pp. 173–174]

The fact that Polish authors truly appreciated measure theory and Lebesgue integrals was greatly significant because it was an age when

> Lebesgue integrals and Lebesgue measure were well known in Poland from 1920 onwards, while in France they had been ignored. [...] One of the books in the Monografie Matematyczne series was devoted to the theory of integrals and was written

[62][C: Ledoux & Talagrand, 1991].

> by Saks[63]. [...] Banach's monograph[64] starts at the very beginning with the assumption that the reader is familiar with measure theory and the Lebesgue integral.
>
> [B: Kahane, 1995; p. 173]

The set of axioms presented by A. Kolmogorov were, naturally, not the only ones possible. B. Gnedenko explained the reason for their success as follows (quotation from [B: Urbanik, 1973; p. 45]):

> They connected powerfully with the spirit of the age, they linked probability theory with the metric theory of functions in a precise manner, and they also fashioned a rich arsenal of research tools, making it possible to embrace, in a single simple framework, not only classical branches of probability theory but also emerging notions and problems as well.

At the same time he added (*ibidem*):

> It is quite apparent that a vibrant mathematical discipline [probability theory], having so many diverse connections with literally every field of knowledge, will always merit detailed research into its foundations, and from many viewpoints. Furthermore, regardless of how well its axiomatic foundations are laid, as long as its development continues so too will the search for justifying it logically in new ways.

As one can imagine, H. Steinhaus was not satisfied with Kolmogorov's solution. One reason might have been because the "Kolmogorovian triple, especially the notion of probability as an abstract normed measure, is not easily reconciled with one's natural intuition regarding practical applications." Moreover, it did not accommodate all the important examples. For example, "the concept of a uniform distribution on a line does not fit in with the Kolmogorovian framework of probability theory, yet it is very important for many problems in physics, statistics, the theory of Markov chains, or probabilistic applications in number theory and analysis" (both quotes taken from [B: Urbanik, 1973; p. 45]). These concepts were formalized by A. Rényi in 1954, but that is another story, one which does not belong to the history of the Lvov school.

H. Steinhaus did not cite A. Kolmogorov in his papers and he did not adopt his axiomatics. Rather, by linking ideas taken from functional analysis and probability theory, he defined stochastically independent functions (similar to independent random variables) and between the years 1936–1940 he published (jointly with M. Kac) six papers under the common title *O funkcjach niezależnych* [On independent functions][65]. The underlying motivation behind them was to "systematically develop an idea contained in Steinhaus's 1923 paper" [A: Kac, 1936]. The third paper in this series presented a mathematical model which simultaneously described the deterministic and probabilitistic motion of a swarm of points in a material (Brownian motions). In the fourth paper he described how to go from Lebesgue measure on

[63]The author has in mind [C: Saks, 1933] (French version) and [C: Saks, 1937] (English version).

[64][A: Banach, 1932].

[65][A: Kac, 1936], [A: Kac & Steinhaus, 1936a], [A: Kac & Steinhaus, 1936b], [A: Kac & Steinhaus, 1938], [A: Kac, 1938], [A: Steinhaus, 1940].

the interval $[0, 1]$ to relative measure on either the whole real line or a half-line. As H. Steinhaus wrote later (quote from [B: Urbanik, 1973; p. 48]):

> This observation leads to a new interpretation of probability calculus; by attaching time as an auxiliary variable one can apply theorems of probability calculus to certain mechanics models, thereby depriving them of a completely random character. One can show [it] in the case of a cube containing a swarm of material particles moving around and bouncing off the walls at equal angles of incidence, in accordance with classical laws. One obtains results for equipartitions of particles, and their center of mass under motion, results that are no different from those obtained via probability calculus. However, the reasoning is entirely different: the classical solution to this problem depends on postulates which cannot be deduced either mathematically or in terms of mechanics, whereas here one relies only on mathematical analysis and the laws of motion in the absence of external forces, and instead of the probability of events one takes their temporal frequencies.

Steinhaus thought about writing a monograph on independent functions, but M. Kac was no longer around, having left for good for the United States in 1938, and those plans got shelved when World War II later broke out[66].

Another of H. Steinhaus's papers dedicated to probability calculus was [A: Steinhaus, 1930a]. In it he introduced the notion of a cylindrical set: it had much significance for product measures.

[66]H. Steinhaus still tinkered with the topic of independent functions because during 1948–1953 there appeared another four works in that series (one of them from the pen of C. Ryll-Nardzewski), but Steinhaus never compiled them into book form.

CHAPTER 16

Measure Theory

The discovery at the end of the ninteenth century of functions having infinitely many special points (points of discontinuity, nondifferentiability, etc.) led to the emergence of Cantor's set theory and measure theory. The main aim underlying Cantor's set theory was the precise description of these special sets (the topology of sets grew out from it), and measure theory made it possible to do integration of certain classes of these singular functions in a natural and intuitive sense (more precisely, in the Riemann sense). Of the several attempts to define the notion of measure, let us mention Jordan measure (1892) and Lebesgue measure [C: Lebesgue, 1902]. The latter is of greater importance than the former because Lebesgue measure is more general than Jordan measure and, moreover, Lebesgue went even further by defining measureable functions (in the Lebesgue sense) and the Lebesgue integral, which is more general than the Riemann integral.

Let us recall that by a measure one means a nonnegative, real-valued function m defined on bounded subsets of the space $\mathbb{R}^n$ and satisfying the following conditions:

(a) $m(I_n) = 1$, where I_n is the n-dimensional unit cube (the nontriviality law);

(b) if A and B are bounded and $A \cong B$ (here $\cong$ stands for congruence), then $m(A) = m(B)$(the invariance condition of measure under rigid movements);

(c) if A and B are bounded, disjoint sets, then $m(A \cup B) = m(A) + m(B)$ (the law of finite additivity);

(d) if $A_1, A_2, \ldots$ are bounded, pairwise disjoint sets and if their union $A_1 \cup A_2 \cup \cdots$ is bounded, then $m(\bigcup_i A_i) = \sum m(A_i)$ (the law of countable additivity; of course condition (d) subsumes condition (c)).

C. Jordan constructed functions m satisfying conditions (a), (b), and (c) for a certain class of functions, but H. Lebesgue constructed functions m for a much wider class of sets, i.e. the so-called measurable (L) sets, which included all the open sets, closed sets, and, even more generally, the Borel sets, which satisfy the stronger conditions (a), (b), and (d). His construction is of major importance today, but at the time it was not clear whether it could accommodate all the bounded subsets of $\mathbb{R}^n$. In 1905 G. Vitali applied the axiom of choice to give an example of a bounded subset of $\mathbb{R}^1$ that is not measurable (L).

In his monograph [C: Hausdorff, 1914; Chapter X, 1], F. Hausdorff restated *the general problem of measure*: construct a function m assigning a real number $m(A)$ to each bounded subset A of the space R_n such that conditions (a)–(d) are satisfied. By modifying Vitali's example he showed that no solution existed to the general problem of measure in the case of $\mathbb{R}^1$ space, and, with that, no solution for $\mathbb{R}^n$ space, for any $n = 1, 2, 3, \ldots$. By making one extra step, he also showed [C: Hausdorff, 1914; pp. 469–472], and this was perhaps the most astonishing result in his monograph, that the measure problem had no solutions in $\mathbb{R}^n$, for $n > 2$, even if one restricts oneself only to conditions (a), (b), and (c).

Hausdorff's proof exploited the fact that the rotation subgroup $SO(3)$ of the three-dimensional Euclidean group contains a free subgroup with two generators. He chose two rotations $\phi^2 = e$, $\psi^2 = e$, where e denotes the identity mapping. Using this he showed that one can represent the unit sphere S^2 as the union of four pairwise disjoint sets Q, A, B, C such that Q is countable and such that there is a congruence relation $A \cong B \cong C \cong (B \cup C)$. If a general measure m did exist, then one would have $m(Q) = 0$ (by the law of countable additivity), and hence $m(S^2) = 3m(A)$. But then, since $m(A) = m(B \cup C)$, it also follows that $m(S^2) = 2m(A)$, which is a contradiction. Therefore the general problem of measure has no solutions in $\mathbb{R}^3$ and, with that, no solutions in $\mathbb{R}^n$ for $n > 2$.

F. Hausdorff's discovery had international impact. But it was still an incomplete result: the general problem of measure remained open for the cases $n = 1, 2$.

When H. Steinhaus and A. Łomnicki were preparing their papers, intending to apply measure theory to probability calculus, the theory must have been a topic of discussion in Lvov (S. Banach recalls S. Ruziewicz coming to him with a problem about it). It was therefore only natural that S. Banach took up the general problem for $n = 1, 2$. He showed—contrary to expectation—that in both of these cases the answer is affirmative [A: Banach, 1923]. S. Banach's solution caused quite a stir because it emphasized a crucial difference between the lower and the higher dimensions. Like Hausdorff's result, it too depended on the axiom of choice, a widely discussed topic at the time, but it bothered neither Hausdorff nor Banach.

It was a period when there were heated discussions about the axiom of choice. Introduced by E. Zermelo in 1908, it states that for every given family of nonempty, pairwise disjoint sets there exists a *selector* set, namely a set which contains exactly one representative element, no more, no less, from each constituent set of the given family. The axiom is very important because many theorems that are basic to mathematics depend on it (for example, in mathematical analysis there is a theorem stating two particular definitions of continuity are equivalent: sequential continuity and $\varepsilon - \delta$ continuity). Moreover, it is very powerful because it leads to various paradoxes such as the Banach-Tarski paradox, which we shall discuss in a moment. Polish mathematicians discussed the axiom at length. W. Sierpiński wrote a long paper [C: Sierpiński, 1918] presenting an analysis of the known results in set theory and mathematical analysis which depend on the axiom of choice. Without committing himself either to accept or reject the axiom, he recognized it was necessary to adopt a neutral stance in such discussions, while clearly marking out, everywhere, all the theorems whose proofs rely essentially on it. The Polish School of Mathematics followed his lead, as did the others around the world [B: Moore, 1982; Section 4.1. *A survey by Sierpiński*, pp. 197–200].

Tarski used to come to Lvov since the early 1920s, and he established a working partnership with Banach. It was not coincidental: Tarski had mastered set theory, while Banach had mastered a nonconstructive method that relied on the axiom of choice, allowing him, for example, to assign a limit to any bounded sequence, whether convergent in the usual sense or not. In their joint work [A: Banach & Tarski, 1924] they produced a truly astounding result, known as the Banach-Tarski paradox: if $n \geq 3$, then any two bounded subsets of $\mathbb{R}^n$ (in particular, two balls of different radii or two arbitrary polyhedra), having nonempty interiors, can be assembled from each other via a finite decomposition:

> If one takes two bodies, A and B, of different volumes and lying in three-dimensional space, they can be broken up into pieces A_j and B_j which can be pasted over each other. It is completely counterintuitive because if we say the A_j and B_j can be pasted over each other, then they must have the same volumes, and therefore the volume of A, being the sum of the volumes A_j, is equal to the volume of B, being the sum of the volumes B_j. Our intuition is both correct and incorrect at the same time: it is correct because if every bounded piece in the space had a volume, then the decomposition, as described, of A and B would be impossible; it is incorrect because it seems obvious that every bounded piece in the space has a volume, but the axiom of choice shows this is actually false.

[B: Kahane, 1995; p. 171]

Later, thanks to the Banach-Tarski paradox, the following special case was made understandable: a ball of radius r can be decomposed into finitely many pieces which can then be reassembled into two balls of radius r. This paradox became famous (see [B: Wagon, 1985]) and one of the most frequently cited paradoxes resulting from the axiom of choice. We add that (as noted by A. Lindenbaum) there is no paradox in dimensions $n = 1, 2$.

Banach's continuing interest in the measure question materialized again a few years later in the following generalization of this problem, having more of a set-theoretic flavor than a geometric one. Specifically, Banach asked himself if there was a function m assigning a nonnegative real number to every subset of the interval $[0, 1]$ while satisfying the following conditions:

(a) $m(E) > 0$ for some subset E,
(b) if E is a point, then $m(E) = 0$,
(c) m is countably additive.

The difference with this and the measure problem considered earlier by F. Hausdorff lies in removing the condition requiring congruent sets to have the same measure—and replacing it with the weaker, set-theoretic condition (b). The motivation behind this change was to ascertain if the existence of nonmeasurable sets (in the Lebesgue sense) on the line, as proved by G. Vitali in 1905, was an unwelcome consequence of requiring measure to remain invariant under translations.

> As I recall, it was sometime around 1928/1929, during one of my many discussions with Banach in the Scottish Café, that we got interested in the so-called measure problem stated many years before by Hausdorff and still unresolved at the time. During a many-hour discussion we tried to attack the problem in various ways, unsuccessfully. I got home around midnight. But I couldn't fall asleep until I had found a solution (to my great joy). I met Banach the following morning. "You know Stefan, I've solved Hausdorff's problem." "So have I," said Banach. What is more, it turns out that my reasoning and Banach's were almost identical: it was essentially a continuation of our discussion in the café. Obviously, we decided to publish it as a joint paper

> (in "Fundamenta")[67]. In that paper we left a certain problem unresolved (and we didn't try to solve it). I told Ulam about it.
>
> Sometime later Ulam came to me with a ready proof. It of course brought me much satisfaction, and because I thought it to be a very valuable result, I encouraged Ulam to write it up in a doctoral thesis[68].
>
> [B: Kuratowski, 1981; p. 95]

In the paper [A: Banach & Kuratowski, 1929], the authors showed that if the continuum hypothesis[69] is true, then it is impossible to assign a measure to all subsets of a set X having the power of the continuum.

In the following year also Banach was continually thinking about the measure problem. He made another contribution to set theory: he replaced the interval $[0, 1]$ with an arbitrary set of power $2^{\aleph_0}$, so that now the problem actually depended on the power of the set to which one wanted to assign a measure. Also, he proposed the following *extended problem of measure*: does there exist a set X (of whatever power) and a function m assigning a number $m(E)$ to every subset E of X such that conditions (a), (b), and (c) are satisfied and which is not identically equal to 0. Assuming the generalized continuum hypothesis, he showed that the power of such a set X cannot be smaller than the first unattainable cardinal number [A: Banach, 1930a; p. 101][70].

Others immediately picked up on Banach's paper. On the one hand, A. Tarski showed that if condition (d), namely countable additivity, is replaced by condition (c), namely finite additivity, in the extended measure problem, then it follows from the axiom of choice that every infinite set has such a measure and that measure can only assume the values 0 or 1 [C: Tarski, 1930]. On the other hand, S. Ulam, inspired by discussions with K. Kuratowski, likewise applied the axiom of choice to show that if every weakly unattainable cardinal number is greater than $2^{\aleph_0}$, then the power of any set satisfying the extended measure problem cannot be smaller than the first strongly unattainable cardinal number [A: Ulam, 1930; p. 150].

Research inspired by S. Banach's work on measure-theoretic problems led to a distinction between measurable cardinal numbers (for a set having such a power, there exists a measure assuming only the values 0 or 1) and real-valued measurable cardinal numbers (for a set having such a power, there exists a measure assuming values different from 0 or 1), and this was incorporated into the very foundations of mathematics; see [B: Moore, 1982].

Banach's investigations into measure theory pulled other mathematicians, such as A. Tarski, K. Kuratowski, and S. Ulam, into its study and had a clear set-theoretical flavor. But it did not have further significance for mathematical analysis, i.e. the framework for which and from which measure theory originally arose. This doubtless explains why H. Steinhaus was not involved in such research: Steinhaus always had a great affinity for mathematics which had a healthy connection to the

[67]This was the paper [A: Banach & Kuratowski, 1929].

[68]That is what happened. The doctoral thesis appeared as a separate 1933 booklet (reprinted in [A: Ulam, 1997]), though the key results appeared earlier in [A: Ulam, 1930].

[69]The continuum hypothesis states that every infinite subset of the real line is equinumerable either with the set of natural numbers or with the whole real line. It does not depend on ZFC (Zermelo-Fraenkel set theory with the axiom of choice).

[70]A cardinal number $\aleph$ is called unattainable if it is a limit number but is not a sum of lesser cardinal numbers, the number of which is less than $\aleph$.

real world, whereas highly abstract, especially set-theoretical, solutions held little attraction for him[71]. On the other hand, the results obtained by S. Banach and his fellow coworkers had a major influence on the general theory of topology.

S. Banach was interested in measure-theoretic questions for the rest of his life. This is clear from his talk at the Lvov branch of the PTM on July 19, 1938, and by posthumously published papers[72].

[71]Such considerations, however, were not unfamiliar to him. See Steinhaus's letter to Ruziewicz, quoted in [B: Więsław, 2004; pp. 166–167].

[72]A list of Banach's posthumously published papers, including the monograph [A: Banach, 1951], as well as the preprints for some of them, are contained in the books [A: Banach, 1967] and [A: Banach, 1979].

CHAPTER 17

Game Theory: A Revelation Without Follow-up

The topic of game theory appeared only once, in a minor, ephemeral academic pamphlet. It was a short work [A: Steinhaus, 1925], the significance of which was surely unappreciated even by the author himself.

> It is a short work, not having the character of a mathematical paper; it essentially consists of several comments, but they are comments which were revelatory for their time, which lie at the very heart of modern game theory. First, the notion of a strategy was introduced most precisely (it was actually called something else, but that is irrelevant). The second important contribution is the so-called normalization of a game and, at last, the notion of a payoff function, which characterizes any given game, as well as the minimax principle of strategy selection.
>
> [B: Ryll-Nardzewski, 1973; p. 29]

The paper was revelatory because game theory was still in its infancy. E. Borel had similar ideas a few years earlier (papers from 1921 and 1924), but they included nothing by way of a theorem about the minimax principle, which was proved by J. von Neumann in 1928. These three mathematicians, however, were completely unaware of each other's work, while H. Steinhaus even had a few different ideas. For example, he came up with the definition of an open game and a closed game (nowadays we speak of deterministic games and nondeterministic games). It is a great pity there was not enough willpower or energy to pursue these issues further. By the time the topic was revived in Wrocław after 1945 (by Andrzej Zięba, under H. Steinhaus's guidance), the whole mathematical scene had changed. Although H. Steinhaus's 1925 paper, after its translation into English in 1960, did become a revelation, it was by then only a revelation from the perspective of historical hindsight.

CHAPTER 18

Operator Theory in the 1920s

Stefan Banach worked intensively, publishing several papers every year (the list of his publications for the years 1919–1940 has fifty-three entries, to which one should add six other works published posthumously in the years 1947–1951; see Figure 22), and his interests were wide, encompassing the theory of real functions, measure theory (see Chapter 16), orthogonal series, etc. An illustration of his versatility can be seen in his earlier paper [A: Banach, 1922b], which includes an example of a curious singularity, prompting further research leading to a more general result, known in the literature as the Fichtenholtz-Banach theorem: for every function f, integrable over any measurable subset E of $\mathbb{R}^k$, there exists a system of orthogonal functions ϕ_n on E such that every function orthogonal to all of the ϕ_n has the form cf, where c is a constant. Upon a suitable choice of the function f, the theorem provides a means of constructing orthogonal systems which are complete in certain function classes containing f but are incomplete in others [C: Fichtenholtz, 1926], [C: Saks, 1931]. Moreover, W. Orlicz worked on a generalization of Banach's idea, in the works [A: Orlicz, 1929–1939; Part I], [A: Orlicz, 1936b]. In the second of these works he showed that in any function space, satisfying certain conditions, there exists an orthonormal system with respect to which the Fourier development of some function in the space is divergent almost everywhere.

The next important joint work of S. Banach and H. Steinhaus was on "the condensation of singularities" and appeared in 1927 [A: Banach & Steinhaus, 1927]. The terminology in the title was first introduced by H. Hankel for labeling a way of constructing an object, having infinitely many singular points, out of infinitely many other objects, each having only one singular point. The number of various theorems, seemingly paradoxical, with proofs exploiting the "gliding hump" method began to multiply from 1910 onwards. However, "no new ideas were forthcoming until 1927, when S. Banach and H. Steinhaus (working in part with S. Saks) established[73] a connection between those results, the notion of a thin set, and Baire's theorem for complete metric spaces [see Chapter 19 for more on this theorem]. By exploiting this connection they obtained a very general result, which included all the previous ones as special cases" [B: Bourbaki, 1980; p. 272]. In other words, S. Banach and H. Steinhaus showed how the cumbersome "gliding hump" method, as applied over the previous two decades, could be replaced by skillful use of the Baire category method (see [B: Dieudonne, 1981; pp. 138–142]). The Banach-Steinhaus theorem for sequences of linear operators turned out to be "one of the fundamental theorems of functional analysis" [B: Orlicz, 1965b].

[73]Of course, this refers to the paper [A: Banach & Steinhaus, 1927].

It was S. Saks who first suggested applying Baire's category method in this context. It is a consequence of this theorem that, for example, the Fourier series of continuous functions is "in general" divergent. Here the expression "in general" actually means this: Baire's theorem says certain sets are "small" (i.e. belong to category I) in relevant spaces of functions, and hence their complementary sets are "large" (i.e. belong to category II, thus are generic). So, with reference to the cited paper, the term "in general" means that divergent Fourier series of continuous functions constitute a set of category II (i.e. a generic set).

The most important paper in the first volume of "Studia Mathematica" is unquestionably [A: Banach, 1929a], on linear functionals. In it S. Banach considered linear and continuous operators (on Banach spaces) and proved many theorems, including generalizations of results of E. I. Fredholm and F. Riesz. Among other things he proved every linear operator is bounded: if U is such an operator and if it has a continuous inverse, then the equation $y = U(x)$ has a solution for every y (this is Banach's inversion theorem for linear operators), together with the extension theorem, later named the Hahn-Banach theorem (see below).

The question of the existence of atypical linear functionals on Banach spaces is of foremost importance in functional analysis. One of the main reasons why the Hahn-Banach theorem is so fundamental is because it provides an affirmative answer to the existence question. Although particular cases were known earlier (see [B: Hochstadt, 1980]), a more general form was given first by H. Hahn [C: Hahn, 1927; p. 217] and then in full generality by S. Banach [A: Banach, 1929a; Th. 1, p. 226]:

The Hahn-Banach Theorem. *Let p be a seminorm on a Banach space X, i.e. a real function satisfying the conditions*

$$p(x+y) \leq p(x) + p(y),$$
$$p(\lambda x) = |\lambda| p(x) \quad \text{for every } x, y \in X, \lambda \in \mathbb{R},$$

and let f be a linear functional defined on a linear subspace E of X such that $|f(x)| \leq p(x)$ for every $x \in E$. *Then there exists a linear functional F on X which is an extension of the functional f*, i.e. $F(x) = f(x)$ for $x \in E$, *and satisfies the condition*

$$|F(x)| \leq p(x)] \quad \text{for every } x \in X.$$

In particular, a bounded linear functional on a subspace of a Banach space can be extended to a linear functional on the whole space without increasing the norm. This follows from the Hahn-Banach theorem, as stated above, by taking $p(x)$ to be a constant multiple of the norm of x. The Hahn-Banach theorem is one of the extension theorems used fairly frequently in mathematics, stating when and under what conditions a function, defined on a subset of a considered space, can be extended to all the space without losing any of its basic properties. More precisely, for a given function $f : A \to Y$, where $A \subset X$, when and under what conditions does there exist a function $F : X \to Y$ defined over all of X, retaining the same basic properties as f, and being identical to f when restricted to the subset A. One of the first such theorems, having much importance to this day, was the subsequent theorem of Tietze: if a space X is normal, then every continuous mapping $f : M \to R$, defined on a closed subset M of the space X and assuming real numbers, has a continuous extension F over all $x \in X$, and furthermore, if the mapping f is bounded, i.e if it satisfies the inequality $|f(x)| \leq c$ for every $x \in M$, then its extension F also satisfies the same inequality for all $x \in X$(see [C: Duda, 1986, p. 142]). Extension theorems are important in mathematics, and the Hahn-Banach theorem in particular is one of the fundamental theorems in functional analysis. It has numerous applications, such as in proofs of generalized limits of functions or sequences (a topic taken up by Banach and Mazur), or for theorems about moments (H. Hahn, S. Banach, S. Mazur, W. Orlicz)[74]. Furthermore, the Hahn-Banach theorem is important for the duality theory of topological vector spaces. N. Bourbaki, being fascinated by the duality theory, devoted nearly an entire book to the theorem [C: Bourbaki, 1953].

Banach's paper introduced a particularly important notion (quite independently of H. Hahn's earlier work, but of even greater importance, as it turned out), namely that of the *dual space* X^* of a Banach space X. The norm of a functional $f \in X^*$ is defined as the supremum of $|f(x)|$ for all $x \in X$ of unit norm. If X is a Banach space, then its dual space X^* is itself a Banach space.

Accompanying the notion of a dual space is the notion of a *dual operator*. Let X, Y be Banach spaces, and let X^*, Y^* be their dual spaces. Then, for each continuous linear operator $T : X \to Y$ there is a corresponding operator $T^* : Y^* \to X^*$ called a *dual operator*, defined by $T^*g = gT$ for $g \in Y^*$.

In [A: Mazur, 1929], S. Mazur proved that the real Banach spaces $L^p[0,1]$ and l^q are homeomorphic, for all $p, q \geq 1$. In his presentation of this theorem at a session of the Lvov branch of the PTM, held on October 28, 1928, S. Mazur asked whether two separable, infinite-dimensional Banach spaces are always homeomorphic. He further stated that every separable, infinite-dimensional Banach space is homeomorphic to some subset of every other space of that type. But he never gave a proof.

Problems such as these motivated research into the topological structure of linear metric spaces and their subspaces. That subsequently developed into a whole research field known as infinite-dimensional topology; see [C: Bessaga & Pełczyński, 1975]. S. Mazur's question was answered by the famous Anderson-Kadec theorem, dating to 1965–1966: every infinite-dimensional Fréchet space (in particular every such Banach space) is homeomorphic to a Hilbert space.

J. Schauder simplified [A: Schauder, 1930a] Banach's proof, given in "Studia Mathematica", vol. 1, of the inversion theorem for linear, bijective operators. He also showed that that theorem leads to a domain-invariance theorem: if a linear, continuous operator is one-to-one (but not necessarily bijective) and if it maps a Banach space to a Banach space, then it maps open sets to open sets.

Continuing his earlier work on fixed point theorems [A: Schauder, 1927], J. Schauder proved Banach's theorem in a more general form. Specifically, he showed that if an operator F, defined on a Banach space, maps a convex, closed set H to itself such that the image $F(H)$ is compact, then it has a fixed point (Satz II in [A: Schauder, 1930b]).

Of the four papers written by S. Mazur in "Studia Mathematica", vol. 2, let us touch on one, in which he showed [A: Mazur, 1930a] that the convex closure of a compact subset of a Banach space is also compact (Mazur's understanding of compactness is what is nowadays referred to as conditional, or relative, compactness). The paper introduced convexity considerations into functional analysis, and

[74]See [A: Mazur & Orlicz, 1953].

its development was further enhanced by later papers of S. Mazur himself (dating from 1933 onwards).

Continuing with his interest in summability theory, Mazur struck upon the excellent idea of applying Banach spaces to summability theory [A: Mazur, 1930b]. From that time onwards methods of functional analysis have dominated the theory of summability to such an extent that the latter is now considered a part of the former; see [C: Zeller & Beekmann, 1965], [C: Wilansky, 1984].

During the years 1928–1938 S. Banach was heavily preoccupied writing academic and school textbooks (see Chapter 9). It was an unusually absorbing undertaking, but fortunately he still had enough time and energy to write a book [A: Banach, 1931c] synthesizing the theory that had grown out from his doctoral thesis. The book was printed in Polish. Its importance was recognized immediately. Already as early as the following year there appeared an expanded version, in French, as the first volume of a new series, Monografie Mathematyzcne [A: Banach, 1932].

While S. Banach was busy writing his monograph [A: Banach, 1932], S. Mazur was ruminating about Banach spaces. In particular, he asked himself whether there always exists a basis for every separable Banach space. He and Banach undoubtedly tried to solve this problem, but without success. They had to content themselves with partial results. For example, they showed that every weak basis in a Banach space is a basis. That theorem is attributed to S. Mazur, but he did not give a proof. The proof was only given later by others in [C: Bessaga & Pełczyński, 1959], in which the authors also generalized Mazur's theorem about Fréchet spaces. There was a big sensation in 1973 when P. Enflö solved the basis existence problem for separable Banach spaces, showing the answer was no (see the commentary to problem number 153 in the *Scottish Book* in [B: Mauldin, 1981] and see Chapter 21).

In a joint theorem (see the monograph [A: Banach, 1932; Chap. XI, 8]) having links to Mazur's earlier 1930 papers, Banach and Mazur proved that every normed separable space is isometrically isomorphic to some subspace of $C[0,1]$. It was the first example of a *universal* space for some class of locally compact spaces. It is worth recalling the beautiful Mazur-Ulam isometry theorem [A: Mazur & Ulam, 1932]: in a normed linear space U over $\mathbb{R}$, which satisfies the linearity condition (namely, if $||x+y|| = ||x|| + ||y||$, then $x = \lambda y$ for some λ), every isometry is an affine mapping.

S. Banach also found unexpected applications of his theory of linear equations of operators [A: Banach, 1930b].

CHAPTER 19

Methodological Audacity

The greatness of the Lvov school can be seen in the intellectual audacity of its founders, H. Steinhaus and S. Banach. The young Steinhaus had graduated *summa cum laude* from Göttingen, considered the world's foremost mathematical center at the time. Having already published his first paper, he audaciously, and successfully, tackled various problems in mathematical analysis. In 1916, while walking in the Planty in Cracow, he overheard the words "Lebesgue integral" being uttered. He reacted immediately, and that is how he met and befriended the slightly younger Stefan Banach (see Chapter 6). One should bear in mind that measure theory and Lebesgue integrals were not widely known back then (this theory appeared for the first time in H. Lebesgue's doctoral thesis, written in 1902, but was received with much criticism; Lebesgue only got his professorship in 1917).

In 1922 Paul Lévy collated and summarized the known and scattered results about function spaces. By that time Stefan Banach had already obtained his doctorate, containing his general definition of type (B) spaces, and had made the first steps in the theory of Banach spaces, as type (B) spaces were called shortly afterwards. Cultivating this theory required knowledge of measure theory and Lebesgue integrals, which were still regarded in France as being of secondary importance. H. Steinhaus immediately began axiomatizing the first nonclassical cases in probability, precisely by exploiting measure theory and Lebesgue integrals. With hindsight one can see that this step, audacious for its time, was a pointer to the axiomatization of probability calculus itself, which would establish it as a mathematical discipline in its own right.

Another example of mathematical audacity from the Lvov School of Mathematics was its approach to nonconstructive methods, namely methods which establish the existence of a mathematical object without specifically constructing it directly, such as proving the existence of continuous, nowhere differentiable functions without actually constructing one. The axiom of choice was a powerful tool in such endeavors. It was formulated in 1908 by Ernest Zermelo and was much discussed during the 1920s by philosophers and mathematicians. This very important axiom led to some paradoxical results, such as the paradoxical decomposition of the ball by Banach-Tarski (see Chapter 16), causing its validity to be questioned. W. Sierpiński's viewpoint prevailed and was adopted by the two Polish mathematical schools in Lvov and Warsaw: the ultimate resolution regarding the validity of the axiom of choice would only be forthcoming with greater knowledge of its consequences, and so therefore no restrictions should be imposed prior to its application.

Baire's theorem of 1899 was another aid for noneffective methods. It states (in modern language) that in a complete metric space, the countable union of nowhere dense sets (in other words, a so-called category I set) is a boundary set. Let us recall that a set is called a boundary set if it contains no open subset of the space (so therefore its complement in the space is dense and is, in particular, nonempty) and a set is called nowhere dense if the interior of its closure is empty. To use Baire's theorem to show the existence of a certain mathematical object, such as a continuous function with some paradoxical property, one showed that all the "well behaved" functions can be defined step by step as elements of nowhere dense sets $F_1, F_2, \ldots$ lying in the considered space. Consequently, there must then be very many "badly behaved" functions, meaning there are a large number of the functions whose existence needed to be established: since they constitute the completion of the union $F_1 \cup F_2 \cup \cdots$, they constitute a dense set in our considered space. S. Banach proceeded eagerly with this line of approach; e.g. in the paper [A: Banach, 1931b] he considered a space C of continuous functions on the interval $[0, 1]$, with the norm

$$\|f\| = \sup_{0 \leq x \leq 1} |f(x)|$$

(it is a Banach space and so, in particular, it is a complete metric space), and its subsets F_{n-1} consist of functions $f \in C$ for which there exists at least one point x belonging to the interval $[0, 1 - 1/n]$ and satisfying the inequality

$$\left| \frac{(f(x+h) - f(x))}{h} \right| \leq n, \quad \text{where } 0 < h \leq 1 - x, \ n = 2, 3, \ldots.$$

The sets F_{n-1} turned out to be closed and nowhere dense, so (by Baire's theorem) their union $F_1 \cup F_2 \cdots$ is a boundary set. Therefore the complementary set in C of this union, consisting of those continuous functions having no right-defined, finite-valued derivative, is nonempty (and dense). In such a manner, without actually constructing any example of such a function, one establishes its existence. In fact, there is more: the functions which have no right-defined, finite-valued derivative anywhere are in the majority, in a categorical sense.

The same theorem was proved (but in a different way) by S. Mazurkiewicz [C: Mazurkiewicz, 1931]. These two papers of S. Banach and S. Mazurkiewicz inspired a series of papers exploiting Baire's category theorem, in various forms, to prove the existence of continuous functions having various curious features, including [A: Auerbach & Banach, 1931], [A: Kaczmarz, 1931], [C: Saks, 1932], [C: Jarnik, 1933], and [C: Jarnik, 1934].

Several illustrative examples of how to apply Baire's category theorem can be found in Volume 3 of "Studia Mathematica", including [A: Banach, 1931b]. We add as a note that the Polish School of Mathematics (equally as much in Warsaw as in Lvov) fashioned Baire's category theorem into an unusually powerful tool [B: Duda, 2008], most particularly for proving existence theorems. The proofs, though nonconstructive, were typically astoundingly short as well as elegant.

> The game plan of the Poles was to choose appropriate topological spaces and describe generic situations. Cumbersome constructions were replaced by existence theorems using Baire's theorem. The topological spaces which work for this kind of game plan were deservedly called Polish spaces by Bourbaki.
>
> [B: Kahane, 1995; p. 172]

One may also apply measure theory and Lebesgue integrals to nonconstructive methods.

The audacity of the Lvov (more generally the Polish) school has its roots in the personalities of its creators. Here is an outline profile of one of the leaders of the Lvov school, as described by another:

> Anyone who believed Banach was a dreamer, an abnegate, an apostle or an ascetic would have been utterly mistaken. He was a realist. He did not even physically resemble someone exuding saintliness, let alone someone faking it just for the sake of the holier-than-thou types. I don't know if there is an idealized image for a Polish scholar nowadays, but twenty-five years ago[75] there certainly was, based not only on the example set by practicing scholars, but also satisfying the spiritual needs of the time, and of which Stefan Żeromski was the greatest personification. Such a scholar was expected to deny himself worldly pleasures for the greater good of some obscurely defined "society", the uselessness of their work being forgiven from the outset. In the meantime, no attention was paid to the fact that in other countries scholars were judged not on the magnitude of their self-denial but on the permanency of their contribution to knowledge. Between the wars[76] the Polish intelligencia still responded favorably to this idealized image of scholastic martyrdom, but that was never Banach. He was healthy and he was strong. He was a realist even to the point of cynicism, yet he gave Polish science, specifically Polish mathematics, more than anyone else. Nobody did more than he to dispel the harmful opinion that a lack of genius (if only a lack of talent) in the academic profession can be substituted by other qualities, particularly those qualities which are difficult to describe. [...]
>
> His greatest service was in once and forever overcoming, and smashing to smithereens, the Polish inferiority complex prevalent in the sciences, which masked its true face by making mediocrity appear superior. Banach himself never suffered from any such complexes—fused together within him were the spark of genius and some sort of amazing inner drive, which perpetually spoke to him as if it were a poem: "There is only this: the ardent glory of the craft!"[77]—and mathematicians well know that their craft withholds the same mysteries as those of poets ....
>
> [B: Steinhaus, 1961a; pp. 258–259]

Steinhaus held this opinion dearly, but to a great extent it applied to him too, when one considers the students and successors who followed on from him.

---

[75]That is to say, in the 1930s.

[76]That is, the two World Wars.

[77]Paul Verlaine: "Il n'y a que la gloire ardente du métier."

CHAPTER 20

Banach's Monograph: Polishing the Pearls

We shall briefly take a look at the contents of Banach's monograph [A: Banach, 1932]. It is dedicated to his wife Łucja (*À Madame Lucie Banach*) and consists of a foreword, introduction, twelve chapters, an appendix, notes, and an index of names—a total of 256 pages.

In the foreword the author expresses his conviction that operator theory, the creation of which he attributes to V. Volterra, blends, in an unusually effective and harmonious manner, classical and modern methods of mathematics, permeating many mathematical fields, including the theory of functions of a real variable, integral equations, the calculus of variations, etc. His book focuses on *linear* operations, especially those defined on type (B) spaces, special cases of which are the space C of continuous functions, the L^p spaces of functions whose p^{th} power is integrable, Hilbert's sequence space ℓ^2, etc. However, that book is only intended as the first part of an even larger work because in the second part he intends to address the remaining functional operators with a heavy emphasis on topological techniques.

The foreword offers a farsighted and promising vision, but this vision seems conditional on historical hindsight. S. Banach still does not see a new mathematical field before him, the foundations of which he himself had already laid down. Moreover, despite a wonderful intuition for dealing with linear operators on type (B) spaces, his primary focus is on operations and not on spaces. It is analogous to thinking about algebra by looking only at homomorphisms instead of looking at groups and rings or thinking about topology by looking at continuous mappings between two spaces only, instead of looking at the spaces themselves. The intended second part was to be devoted to nonlinear operators, a field which had been opened up more widely by J. Schauder and his topological methods, and although even here Banach's intuition did not desert him—nonlinear functional analysis has become increasingly important ever since—the second part just never was written.

The introduction consists of two parts. It begins with the statement which struck J.-P. Kahane's attention: "We assume the reader is familiar with measure theory and Lebesgue integrals" [B: Kahane, 1995; p. 173], after which the author recalls a few further notions and theorems regarding that theory. In the second part there are reminders about metric spaces (i.e. spaces equipped with a metric—see page 87 of this book for a definition of a metric; the author names them type (D) spaces), specific metrics for specific kinds of function spaces, together with topological notions such as completeness, Baire categories, etc.

Chapter I of Banach's book introduces groups. A group having a metric and such that the group action is continuous (in other words, a topological group) is named a type (G) space by Banach. We add that topological groups, introduced in

1925 by F. Leja and O. Schreier, were little known novelties at the time the monograph was written, but Banach was nonetheless aware of them and fully appreciated them.

Chapter II is devoted to vector spaces, which Banach calls type (D) spaces (he implicitly assumes, without explicitly saying so, that they are vector spaces over the field of real numbers). And right after the definition of an additive operator (linear functional) we have the extension theorem, given in [A: Banach, 1929a], with a short proof, and now known as the Hahn-Banach theorem (see Chapter 18). The fragment nicely illustrates Banach's characteristic style: clear language, simple definitions and proofs, and good examples.

In Chapter III he imposes metric and completeness conditions on a vector space X, requiring that the metric be invariant with respect to translations and that the product λx be continuous with respect to a real-valued scalar variable λ for every $x \in X$, as well as continuous with respect to the vector variable x, for every $\lambda \in \mathbb{R}$. This is how type (F) spaces[78] arise, and they obviously form a subclass of type (D) spaces, but the author shows they also form a subclass of type (G) spaces. What then follows is one of the most astonishing results in the whole monograph, a massive generalization of the Banach-Steinhaus theorem, using Baire category methods (Theorem 3): if $u : E \to F$ is a continuous linear operator from a normed complete space E to a normed complete space F, then either $u(E)$ is a type I Baire category (meager) or $u(E) = F$. One of its immediate consequences is the following theorem about the closed graph: if $u : E \to F$ is a linear operator and if its graph is closed in $E \times F$, then u is continuous. The latter theorem "turned out to be of primary importance in the new functional analysis" [B: Bourbaki, 1980; p. 272]. Its significance is illustrated by a short, elegant proof given at the end of the chapter, showing there exists a continuous function having no derivatives on sets of positive measure and also having some applications to differential equations, systems of linear equations in infinitely many variables, etc.

The proper treatment of the theory of type (B) spaces starts in Chapter IV. It commences with the notion of a *norm*. A norm makes it possible to define a *metric*. If the vector space is equipped with a norm and is furthermore complete with respect to the assigned norm, then he calls it a *type (B) space*[79]. Every type (B) space is a type (F) space, but the converse is false. To ensure that the theory of type (B) spaces was not vacuous, the author addresses the question of whether there exist nontrivial functionals or if functionals defined on subsets of the space can be extended to the whole space. He cites his own results but also makes reference to those of H. Hahn, F. Riesz, and E. Helly. Quite a large part of that chapter is taken up by theorems concerning the form of general linear functionals on some of the basic spaces, such as C, L^p, c, ℓ^p, m, etc., where he cites and generalizes results of F. Riesz, M. Fréchet, and H. Steinhaus (most of the results and transparently clear formulations nonetheless came from S. Banach).

The chapters that then follow introduce notions specific to type (B) spaces: they start with totally continuous adjoint operators and dual operators (Chapter IV) and bi-orthogonal sequences and bases (Chapter VII). The notion of a basis

[78]The name almost certainly stood for M. Fréchet, the first person to take any notice of such spaces.

[79]Although the spaces introduced by Banach, in the doctoral thesis he had written ten years before, were already known as "Banach spaces", he himself always stuck to his own terminology for the rest of his life.

of type (B) spaces was first introduced by J. Schauder [A: Schauder, 1927] and his influence is evident everywhere here[80]. In Chapter VII we find a question which would later be much discussed: does every separable type (B) space have a basis? Further on in the monograph we have linear functionals on type (B) spaces, weak convergence of functionals (Chapter VIII), weakly convergent sequences of elements (Chapter IX, citing results of F. Riesz and J. Radon), equations of linear functionals (Chapter X, general and simpler formulations of results of F. Riesz, T. H. Hildenbrandt, and J. Schauder, as well as integral equations of M. Fréchet and V. Volterra), isometry, equivalence, isomorphism (Chapter XI, uniform and simpler formulations of earlier results of S. Mazur, S. Ulam, K. Borsuk, and others), linear dimension (Chapter XII).

There are two different notions of weak convergence in type (B) spaces, namely weak convergence of linear functions and weak convergence of elements. A few more theorems on these notions are added in the appendix of Banach's book.

An important part of the monograph is the Remarks (*Remarques*), substantial editorial assistance being provided by S. Mazur (S. Banach thanks him for it in the *Foreword*). They include historical comments, remarks about the links between different notions and theorems, open questions, and pointers for pursuing further research.

The notions and theorems given in Banach's monograph turned out to be equally important for functional analysis itself, as well as for its applications, as shown by the author himself in numerous examples. The monograph quickly became widely recognized, so that its subsequent impact on the development of functional analysis was similar to the decisive influence on the development of algebra played by van der Waerden's *Algebra*, written in 1930. Banach's notation and terminology were commonly adopted, while normed, complete vector spaces widely became known as *Banach spaces.* Moreover, functional analysis has since been thought of as "the study of topological vector spaces and mappings $u : \Omega \to F$ from a part Ω of a topological vector space E into a topological vector space F, these mappings being assumed to satisfy various algebraic and topological conditions" [B: Dieudonné, 1981; p. 143].

Banach's monograph firmly established the importance of the Lvov school and considerably enhanced the development of functional analysis around the world, in the direction set out by that school.

> The appearance of Banach's treatise on "linear operations" heralds, shall we say, the beginning of the mature age of the theory of normed spaces. All the results are [...] interspersed with many suitable examples, taken from various analysis fields, that bode extremely well for the future of the theory. The work was immensely successful, and one of its direct consequences was the almost total adoption of the terminology and symbolism used by Banach.
>
> [B: Bourbaki, 1980; p. 272].

[80]Nowadays the basis defined in the way he understood it is called a Schauder basis. From the closed graph theorem it follows that on Banach spaces every Schauder basis is a basis in the usual sense, but for the more general class of locally convex spaces these notions are different. See [C: Albiac & Kalton, 2006].

> One can hardly exaggerate the influence that Banach's book has had on the development of functional analysis. Embracing a much wider field of mathematical questions that is provided by Hilbert space theory, it has probably stimulated a greater volume of works than Stone's and von Neumann's books[81] combined. Because of its greater generality, moreover, the theory of Banach spaces has retained much more of the original flavor of functional analysis (as anticipated or interpreted by Volterra, Frechet, and F. Riesz) than the theory of linear operators on Hilbert spaces.
>
> The book became quickly accepted as the climax of a long series of works initiated by Volterra, Hademard, Fréchet, and F. Riesz. To those liking generality, it could be said to contain much of Hilbert space theory, including the spectral theory of compact operators, as a special case. Thus it acquired an impressive and substantial theory of its own. This went far to establishing functional analysis as a broad and independent field of research.
>
> [B: Birkhoff & Kreyszig, 1984; p. 315]

> It is undoubtedly one of those books that has maximal impact on contemporary mathematics. Although the theories developed in the book [...] could have exploited methods developed previously for more specific goals, which the authors, including Frederik Riesz, foresaw, [...], their overall effectiveness was created almost entirely by Banach and his coworkers. Frederik Riesz always talked about the value of the book with the greatest respect, which he articulated in the review of *Teoria operacji*, written for a Szeged journal, "Acta Scientiarum Mathematicum".
>
> [B: Szökelfalvi-Nagy, 1961]

> His [Banach's] achievements are characterized above all by a masterly exploitation of topological methods, with the aim of obtaining deep theorems in functional analysis, which had not been noticed by his predecessors or his contemporaries. In his famous monograph Banach presented his ideas in a mature and concise manner, emphasizing, with unusual clarity, subtle interdependencies between algebraic and topological considerations, while devising really fruitful abstract and general notions for contemporary functional analysis. What made Banach's results so powerful was the way they unified a series of diverse, very scattered, and incomplete results in analysis.
>
> [B: Stone, 1961; p. 266]

S. Banach was still active in research after writing his monograph, but a large part of his time was taken up writing academic books and school textbooks (Figure 22).

---

[81]The authors almost certainly have in mind the books [C: von Neumann, 1932] and [C: Stone, 1932], which sparked the rapid development of Hilbert space theory.

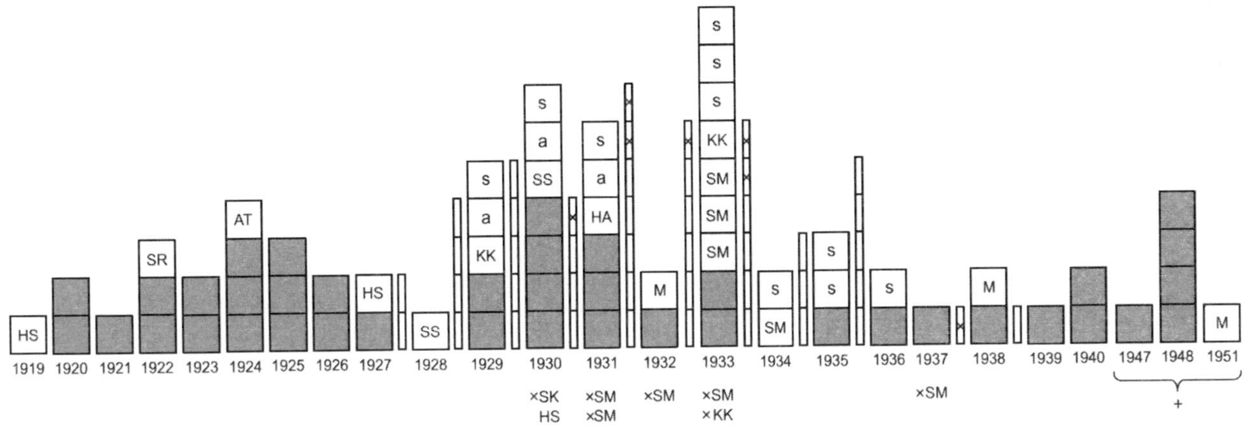

FIGURE 22. Banach's output in numbers. Every square corresponds to one paper. A shaded square indicates sole authorship, an unshaded one denotes a joint paper (coauthors: HS—Hugo Steinhaus, SR—S. Ruziewicz, AT—Alfred Tarski, SS—S. Saks, KK—K. Kuratowski, HA—H. Auerbach, SM—S. Mazur), [a] an academic textbook, [s] a school textbook (eight out of every ten were coauthored by Banach), [M] a monograph. Works from 1947, 1948, and 1951 appeared posthumously. The thin columns appearing to the immediate right of the squares for the years 1927–1938 denote communications delivered in sessions of the Lvov branch of the PTM. An empty thin box means a communication delivered personally, whereas a thin box with cross denotes a joint communication (joint contributors were SK—S. Kaczmarz as well as HS, SM, and KK).

CHAPTER 21

Operator Theory in the 1930s: The Dazzle of Pearls

The definition of a general "type (B)" space [A: Banach, 1922a] was an eminently good one, amply validated by results obtained by the Lvov school during the 1920s (see Chapter 18) and by the success of Banach's monograph, synthesizing the known results of the day (see Chapter 20). This served as a solid basis for the subsequent development of the theory of Banach spaces and, more generally, of functional analysis. Progress would continue in Lvov, due to the master himself (S. Banach and Steinhaus would be pursuing different interests by that time) and equally due to the new acolytes (H. Auerbach, S. Kaczmarz, S. Mazur, J. Schauder, S. Ulam, and others), as well as around the world, particularly in the Soviet Union, France, and the United States. Although the Lvov school would continue to play an outstanding role in the development of the theory, the importance of other foreign centers was, nevertheless, to rise rapidly. There was no lack of talent in Lvov. What was bad was the scarcity of paid positions of employment. In addition, many people were worried about the increasingly tense political situation and the ever greater threat of war, prompting some of them to emigrate (notably Z. W. Birnbaum, M. Kac, and S. Ulam).

The theory of Banach spaces initiated and substantially developed in Lvov is now referred to as classical Banach space theory. Its next golden age would be the period 1950–1980, when Polish mathematicians would also make significant contributions. A summary of this particular period appears in the book [C: Lindenstrauss & Tzafriri, 1977] and in [C: Lindenstrauss & Tzafriri, 1979] (updated in [C: Lindenstrauss & Tzafriri, 1996]). The theory continues to unfold, though nowadays the principal focus has moved on to the general theory of Banach spaces and to more general spaces, most particularly locally convex spaces.

We shall now describe some of the typically Lvovian threads that entwine the very fabric of classical Banach space theory. Although an idea for L^{Φ} spaces had already appeared earlier, in the paper [A: Birnbaum & Orlicz, 1931], the true theory of these spaces is considered to begin with the paper [A: Orlicz , 1932], where they were defined and investigated in a systematic way. The L^{Φ} spaces extend the notion of L^p spaces, the notion of integrability of the pth power of a function being replaced with the notion of integrability of a superposition of functions with a given convex function Φ having certain properties. Later on, L^{Φ} spaces became known as *Orlicz spaces*. The first synthesis of this material in the form of lecture notes was given in the monograph [C: Zygmund, 1935]. These spaces gained wider popularity thanks to the work [C: Krasnosielskij & Rutickij, 1961], the works of A. W. Luxemburg and A. Zaanen, and also because of W. Orlicz's school in Poznań.

An interest in summability theory led S. Mazur and W. Orlicz to a generalization of Banach spaces in the form of the so-called B_0 type spaces [A: Mazur & Orlicz, 1933]. To this end, they first defined a space B_0^* (which, in the terminology of N. Bourbaki, is now commonly referred to as a linear topological, metrized, locally convex space). If such a space is also complete, then it is called a type B_0 space (in Bourbakian terminology, it is a Fréchet space[82]). In his habilitation thesis [A: Mazur, 1936], S. Mazur introduced type B_α spaces (the α stands for the power of a minimal family of pseudonorms describing the topology). S. Mazur and W. Orlicz went a long way in developing the theory of these spaces, but unfortunately they published their results only after the war had ended [A: Mazur & Orlicz, 1948], [A: Mazur & Orlicz, 1953].

Continuing with his earlier interest in convexity [A: Mazur, 1930a] and following H. Minkowski, S. Mazur introduced, in [A: Mazur, 1933a] and [A: Mazur, 1933b], the notions of hyperplane, linear manifold, convex body, etc. There he proved, among other things, that if C is a convex body in a normed space E and if M is a linear manifold containing no interior point of C, then there exists a closed hyperplane H containing M such that C lies on one side of H. Both of those 1933 papers of S. Mazur kick-started intensive research into infinite-dimensional convex sets and the geometry of the unit ball in Banach spaces.

S. Mazur and W. Orlicz generalized, in [A: Mazur & Orlicz, 1933], H. Hahn's theorem on moments, and they introduced a boundedness notion (for Fréchet spaces).

W. Orlicz introduced the notion of unconditional convergence [A: Orlicz, 1933].

While they were following (unwittingly) in the footsteps of S. Mazur and W. Orlicz, in 1934 A. Kolmogorov and J. von Neumann, working independently of each other, introduced the notion of a linear topological space. They also gave definitions for a locally convex space, imposing certain constraints, but they named them in different ways. The modern definition of a locally convex space was given in [C: Tychonoff, 1935]: it is a linear topological space in which every neighborhood contains a convex neighborhood. In that same paper, Tychonoff generalized J.Schauder's famous fixed point theorem to the case of real-valued, locally convex spaces: if C is a compact, convex subset of such a space, then every continuous map sending C to itself contains a fixed point.

Continuing further with the convexity theme, M. Eidelheit generalized Mazur's 1933 theorem about the separation of a convex set by a hyperplane, showing that two convex bodies, having no common interior points, can be separated by a closed hyperplane [A: Eidelheit, 1936a].

Problem number 153 in the *Scottish Book* later became famous. Posed by S. Mazur on November 6, 1936, it promised a "live goose" by way of reward for a solution. Despite its seemingly innocuous technical formulation, the true importance of the problem grew ever more apparent as the years passed by: other results, later on, showed it actually boiled down to proving/disproving the existence of a basis in a Banach space (more precisely, the existence/nonexistence of Schauder bases in separable Banach spaces).

---

[82]One can find references in the literature (see [C: Żelazko, 1968], [C: Żelazko, 1973]) in which a Fréchet space means something else, namely, a linear topological space which is metrizable in a convex sense. If, furthermore, the space is locally convex as well, then one ends up with a type B_0 space in the sense of Mazur and Orlicz.

> Given a continuous function f defined on the closed unit square $[0,1]^2$ and given a positive real-number ϵ, the task was to ascertain if there exist numbers $a_1, \ldots, a_n, b_1, \ldots, b_n, c_1, \ldots, c_n$ such that for all points (x, y) in the unit square, the following inequality holds:
>
> $$\left| f(x,y) - \sum_{k=1}^{n} c_k f(a_k, y) f(x, b_k) \right| \leq \epsilon.$$
>
> It was not until 1955 when Alexandre Grothendieck (Fields Medallist, 1966) showed that the purported existence of such numbers was equivalent to saying every compact linear operator $T : X \to Y$, where X, Y are Banach spaces, is the limit (with respect to the norm) of a sequence of finite-rank operators. A space Y satisfying the above condition for arbitrary X and T is said to have the "approximation property". It was well known that a space would satisfy this property if it had a Schauder basis—and so now at last we arrive at the modern formulation of the problem.
>
> Per Enflö proved that a separable Banach space does not necessarily have a Schauder basis and that not every compact linear operator is the limit of some suitable sequence of operators. In so doing, and by virtue of Grothendieck's result, he had shown that the answer to Mazur's question is negative[83].
>
> [B: Ciesielski, 1993a; p. 67]

The sagacity exemplified by S. Banach and the Lvov school of functional analysis is evident in the role the school played in the formulation and proof of what N. Dunford and J. Schwartz called, in their excellent monograph [C: Dunford & Schwartz, 1958], the three basic principles of functional analysis. The first of them is known as the Uniform Boundedness Principle.

Uniform Boundedness Principle. *Let $T_a : X \to Y$ be, for each $a \in A$, a continuous linear operator from a Banach space X to a normed vector space Y. If the set $\{T_a x : a \in A\}$ is bounded for each $x \in X$, then this set is uniformly bounded for all $a \in A$ and all $x \in X$ of prescribed norm.*

The significance of this principle is that it makes it possible to dispense with two separate assertions, in which one of two parameters is fixed while the other varies, in favor of a single assertion in which both parameters can be varied simultaneously. Many fundamental theorems about linear spaces follow from it.

The principle was first proved by H. Hahn [C: Hahn, 1922]. However, S. Banach and H. Steinhaus noticed that it could be proved under a weaker assumption and, moreover, they proved associated theorems concerning the condensation of singularities [A: Banach and Steinhaus, 1927]. Furthermore, S. Banach derived from the Uniform Boundedness Principle two important and frequently applied convergence theorems [A: Banach, 1922a] and produced related theorems regarding complete metric groups [A: Banach, 1931a]. Other generalizations were investigated

---

[83] News of the result caused a sensation and its author was invited to go to Warsaw, where S. Mazur personally handed him the promised prize.

by S. Mazur and W. Orlicz ([A: Orlicz, 1929–1939], [A: Mazur & Orlicz, 1933], [A: Mazur & Orlicz, 1934]) and also, after the war, by A. Alexiewicz [C: Alexiewicz, 1950–1952].

The following theorem of Brouwer on *domain invariance* is of fundamental importance for the topology of Euclidean spaces (and even more generally, finite-dimensional, separable metrizable manifolds): if A is an open subset of the Euclidean space $\mathbb{R}^n$ and if $h : A \to \mathbb{R}^n$ is a homeomorphism, then the image $h(A)$ is also open in $\mathbb{R}^n$ (see [C: Duda, 1986; p. 353]). If, however, one replaces $\mathbb{R}^n$ with a Banach space, then the above theorem is no longer valid, because there exist homeomorphisms (even linear ones) which map a Banach space to one of its sparse subsets, meaning that no open set is mapped to an open set. It might therefore have seemed as though topology could not offer anything particularly useful for Banach spaces. However, such is not the case, by virtue of a particular theorem about these spaces, stating that if an operator is linear, continuous, and "onto", then it sends open sets to open sets. N. Dunford and J. Schwartz called it the second basic principle of functional analysis, known today as *the open mapping theorem*, meaning that open sets are always mapped to open sets. S. Banach was the first to give a proof for it [A: Banach, 1929a; Th. 7, p. 238]. Subject to some additional assumptions, he proved that a linear, continuous, and bijective operator has an inverse operator which is also linear and continuous. J. Schauder also did work on this principle but did not assume that the operator was linear ([A: Schauder, 1929], [A: Schauder, 1930a], [A: Schauder, 1932a]).

The problem of ensuring that open sets are mapped to open sets has two equivalent reformulations. The *inverse mapping theorem* says that if $T : X \to Y$ is a linear, bijective, continuous operator between Banach spaces, then its inverse $T^{-1} : Y \to X$ is a linear, continuous operator. By contrast, the *closed graph theorem* states that a linear operator from a Banach space to a Banach space must be continuous if its graph is closed. These are basic results and are therefore frequently applied in theorems about Banach space operators (and important analogous adaptations awaited with regard to other kinds of vector spaces).

The third principle of functional analysis addresses the question of when linear functionals can be extended. It is called the Hahn-Banach theorem (see Chapter 18). The first extension theorem was proved by H. Hahn [C: Hahn, 1927; p. 217], but S. Banach proved it in greater generality [A: Banach, 1929a; Th. 1, p. 226] and applied it systematically.

CHAPTER 22

New Perspectives for Which Time Did Not Allow

Although the fame of the Lvov School of Mathematics principally rests with the classical theory of Banach spaces, and also with measure theory to some extent, other mathematical fields were certainly represented as well and were later developed more fully, only elsewhere. We have already mentioned game theory (Chapter 17), so here we shall describe topics which were considered in Lvov in the 1930s, but not developed there more fully due to a lack of time or energy.

22.1. Nonlinear functional analysis

Juliusz Schauder had enormous mathematical talent, but that talent was denied the opportunity to mature in full. Not only were his great abilities already evident in his earliest papers, so too was his deep erudition and audacity in overcoming intellectual obstacles.

Under S. Banach's guidance, some of his early papers were devoted to the theory of integrals. Among other achievements, he made Stokes's equation more precise, introducing topological themes into his arguments. He obtained a doctorate in this field in 1924 on the basis of the work [A: Schauder, 1926], after which he moved on to consider the topology of Banach spaces. Specifically, he worked on Brouwer's fixed point theorem, generalizing it to the case of Banach spaces with basis ([A: Schauder, 1927], habilitation thesis). He also worked on the question of open sets and investigated the conditions under which maps defined on Banach spaces (not necessarily linear maps!) would send open sets to open sets ([A: Schauder, 1929], [A: Schauder, 1930a], [A: Schauder, 1932a]). In the first of these three papers, the author started with the observation that there exist continuous, bijective, linear maps from Hilbert space to itself such that open sets are mapped to nowhere dense sets. In light of this result, one can see that the topology of infinite-dimensional spaces differs radically from the topology of Euclidean spaces. Consequently, the question of what happens to open sets belonging to the former of these spaces needs to be treated in an entirely different way. By abandoning the linearity assumption, J. Schauder assumed that a given Banach space X has a basis and is weakly compact and that the map $U : X \to X$ has the property that $U(x) - x : X \to X$ is totally continuous. Under these assumptions, U maps open sets to open sets. The author broke his proof up into two parts: a topological part in which he generalized L. E. J. Brouwer's proof of the area invariance theorem and an analytical part. He immediately demonstrated the power of his result: he proved, under some additional conditions, that if a second order, elliptic partial differential equation has a solution, then there exists another solution in a neighborhood of the first solution. And so, the following appeared quite early on in Schauder's creative output: the theory of integrals, functional analysis, and topology and the theory of elliptic partial differential equations. He is regarded as a pioneer in the field of nonlinear functional analysis, a subject to which Banach himself placed much importance, intending to write a second volume to accompany his first monograph, but he never managed to write it.

In the academic year 1932/1933 Schauder accepted a two-year Rockefeller grant. He was briefly in Leipzig (staying with L. Lichtenstein), but he spent most of that time in Paris. There he met J. Leray, who had read his paper [A: Schauder, 1932a] and was immensely impressed by Schauder's topological methods. In fact, he considered them to be so powerful that they ought to lead to a theorem about the existence of solutions to the type of equations stated in the title of the paper, without any assumptions about the uniqueness of the solutions. Together they succeeded in proving the theorem, and they then gave a beautiful illustration of how it could be applied; namely they showed the existence of a solution to a problem stated by Dirichlet, defined everywhere over a flat and convex area [A: Leray & Schauder, 1934]. It earned them the 1938 International Melaxa Prize, awarded for the most prestigious mathematical achievements.

J. Schauder defined [A: Leray & Schauder, 1934] the *topological degree* of a mapping $y = \Phi(x) := x - F(x)$, defined on a Banach space, where F is a totally continuous mapping. If D is an area and D' is its boundary, then the topological degree is defined for $y = \Phi(x)$ lying outside D'. It remains invariant under continuous changes and is zero when y does not belong to $\Phi(D)$. If $\Phi(D)$ is a one-to-one mapping and if $y \in \Phi(D)$, then the topological degree is either 1 or -1. For $y = 0$, the topological degree is called an *integral indicator* of the fixed points of the mapping $F(x)$: it is an additive function, it assumes integer values, and continuous modification does not change its value. For example, the indicator of a zero of an analytic function $\Phi(x)$ of a complex variable x is the order of multiplicity of that zero and is strictly positive. These properties of the degree function made it possible to establish the existence, and in some cases even the uniqueness, of solutions to various differential equations in analysis.

It was a significant achievement, not only because it benefited analysis itself, but also because it demonstrated that deep topological methods could be successfully applied to functional analysis, even though this had at first seemed impossible. This achievement marked the start of the algebraic topology of Banach spaces.

Schauder and Leray never succeeded in writing a second joint paper, though that had been their intention. However, they did influence each other greatly through postal correspondence and personal meetings (such as at Geneva in 1935 and also in Warsaw and Lvov in 1938). J. Leray recalled that J. Schauder did not publish that he had succeeded in extending the theory of the algebraic degree for vector spaces having compact neighborhoods [B: Leray, 1980].

S. Banach understood and appreciated the significance of nonlinear operators, as evidenced in the introduction to his monograph [A: Banach, 1932], where he writes that following on from the "theory of linear operators" the second volume would be dedicated to the "theory of nonlinear operators". There was still some time remaining. S. Mazur and W. Orlicz [A: Mazur & Orlicz, 1934], as well as S. Banach himself [A: Banach, 1938a] wrote papers on polynomial operators, in other words, operators belonging to a more general class than linear ones, and J. Schauder was also on hand to help out, but S. Banach never even got round to putting pen to paper with regard to the second volume. Modern nonlinear functional analysis, though less well developed than the linear case, is gaining ever greater importance.

According to J. Leray (see his introduction to [A: Schauder, 1978; p. 15]), Schauder's greatest achievement was to "fonder la topologie algébrique des espaces de Banach, réduire les problèmes classiques de la théorie des équations aux derivées partielles à la preuve que certaines applications linéaires d'espaces fonctionnels ont une norme finie"[84].

22.2. The theory of partial differential equations

The great contribution made to mathematics by the school of differential equations set up in Cracow by Tadeusz Ważewski was its emphasis on the application of topological methods (the so-called Ważewski *topological principle*; see Chapter 28, year 1947) to this theory, and it would later assume much significance after World War II. In fact, the subject enjoyed an even more stunning start in Lvov. J. Schauder, who was interested in partial differential equations, began writing a paper in which he linked problems involving these equations (and integral equations) with a more general approach exploiting functional analysis and using the topological methods he himself had devised for Banach spaces. One of his most intriguing results concerns a second-order, elliptic partial differential equation. He showed that a certain problem of Dirichlet regarding this kind of equation, defined on a convex contour, always has a solution (announced in [A: Schauder, 1932b], result given in [A: Schauder, 1933], extension given in [A: Leray & Schauder, 1934]). It was a major improvement on earlier results of S. Bernstein (who worked with series representations and who had to assume the coefficients did not depend on a function variable). Even though Schauder showed some other nice applications of his method [A: Schauder, 1934a], [A: Schauder, 1934b], it was his improvement of the general case which pleased him most. He further concentrated on the linear Dirichlet problem, devoting seven papers to it in total, including [A: Schauder, 1936b].

> Then Schauder publishes in [Schauder, 1934a] the first version of his method; a year later he writes another in [A: Schauder, 1935b]—an uncommonly elegant and short work. The nine pages in [A: Schauder, 1935b] and the six pages making up Chapter IV of [A: Schauder, 1934b] constitute a complete theory for the linear Dirichlet problem. The note [A: Schauder, 1934c] gives an important variant. That theory is astounding in its simplicity and its penetrative power.
>
> [B: Leray; 1959; p. 18]

The author presented the heart of his methods in [A: Schauder, 1936c], and he foresaw the possibility of extending it to elliptic partial differential equations of arbitrary order. His predictions were proved correct by M. I. Wiszyk (1951), L. Gårding (1951), and F. E. Browder (1951–1956), but J. Schauder had long been dead by then.

Second-order, hyperbolic partial differential equations became a research topic because of J. Hadamard's treatise of 1932. In it Hadamard provided a relatively easy solution to Cauchy's problem for the case of linear equations having constant coefficients but gave a complicated solution for the variable coefficient case, one which

[84] "The creation of an algebraic topology on Banach spaces, thereby reducing classical problems in the theory of partial differential equations to the problem of proving whether or not certain linear mappings of function spaces have finite norm."

was neither suitable for investigating the properties of the solution nor good enough to solve the nonlinear case or mixed problems of Cauchy-Dirichlet type. J. Schauder came up with a new method, analogous to his earlier work on the Dirichlet case, which accommodated both the linear and the nonlinear case [A: Schauder, 1935a], [A: Schauder, 1936a], as well as problems of mixed type [A: Krzyżański & Schauder, 1936], [A: Schauder, 1936a]. He summarized it in [A: Schauder, 1936c], and in his last paper [A: Schauder, 1937] he gave an elegant variant. I. Pietrowski generalized his method to hyperelliptic equations of arbitrary order.

22.3. The theory of linear topological spaces

Functional analysis is not confined to the classical theory of Banach spaces. It accommodates more general spaces equally well as more specific ones. An example of generalization was the change from normed linear spaces to linear topological spaces, to which N. Bourbaki devoted a whole monograph [C: Bourbaki, 1953]. One of the obstacles to finding generalizations was the possibility that the pointwise convergence of a function sequence, defined on the unit interval $[0, 1]$, might not be expressible in terms of a metric; specifically, there might not exist a metric in $C[0, 1]$ necessary to ensure the convergence in the sense that that metric is equivalent to convergence in the pointwise sense. However, it is possible to ensure that these two types of convergence are equivalent if the metric is replaced with a suitable topology, specifically a linear topology, in which pointwise convergence can be characterized by a system of neighborhoods. Unfortunately, this generalization has the drawback that it does not allow for the construction of a duality theory. To try to overcome this drawback, J. von Neumann and A. Kolmogorov, working independently, introduced the notion of a locally convex linear topological space, subject to some restrictions, while naming it differently [C: von Neumann, 1935].

It is a curious thing that S. Mazur arrived at the notion of a locally convex space by himself, independently of J. von Neumann and A. Kolmogorov, and he definitely spoke about it on February 9, 1935, before an assembly of the Lvov branch of the PTM (see his footnote in [A: Mazur, 1936]). At the same time, he gave many results about these spaces, some of them without proof (which were later completed by G. Birkhoff and S. Kakutani; see [B: Köthe, 1994; pp. 227–228]).

The notion of local convexity, taken up and developed by N. Bourbaki, led to the theory of distributions after 1950. The first step in that direction was taken by O. Nikodym, who published a paper in 1933 that included the concept of what would later be known as Sobolev spaces. The paper went unnoticed, and he himself did not pursue it further, and the priority claim belongs to S. L. Sobolev, who announced his "generalized functions" in the years 1936–1937. However, a clear form of the theory was not forthcoming until L. Schwartz defined distributions as linear functionals, continuous on the space of infinitely differentiable functions and having compact supports [C: Schwartz, 1950–1951]. Such a space is a locally convex, but nonmetrizeable, linear topological space. From this perspective one can justifiably say that after 1935 the theory of Banach spaces became part of the more general theory of locally convex spaces, but nonetheless it did not lose its importance, and nowadays there is even revived interest in the geometry of Banach spaces and in other aspects of their particular properties; see [C: Johnson & Lindenstrauss, 2001–2003].

Hilbert spaces are an example of spaces that are narrower than Banach spaces. Their axiomatic definition was given by J. von Neumann in 1929, who published a monograph [C: von Neumann, 1932] about it shortly afterwards. The monograph became the foundation for applications of Hilbert spaces to quantum theory; specifically, it established the principal mathematical apparatus for that theory.

22.4. The theory of recursive functions

The question of computability appeared in Lvov at almost exactly the same time as the appearance of the fundamental work on the subject [C: Turing, 1936].

> But it was in Poland, before World War II, where Banach and Mazur developed this idea in the most systematic way. The war prevented them from publishing the work they had done during that period, so that only an abstract [A: Banach & Mazur, 1937] remained.
>
> [B: Guillaume, 1994; p. 288]

A monograph on the subject appeared after the war [C: Mazur, 1963], produced by A. Grzegorczyk and H. Rasiowa on the basis of notes taken from S. Mazur's lectures, given in the academic year 1949/1950, but it contains nothing new in a historical sense.

22.5. Algebra, specifically Banach algebras

Algebraic arguments appeared in the papers [A: Żyliński, 1932], [A: Eidelheit, 1936b], [A: Eidelheit, 1938], but it was S. Mazur who first took an intense interest in the subject. He noticed more and more algebra in Banach spaces and appreciated its significance for the theory of these spaces. Unfortunately this was also accompanied by an ever greater reluctance to publish his findings. All the same, he did publish a short paper in 1938 [A: Mazur, 1938] in which he introduced a ring equipped with a norm. It states a theorem about the structure of a normed field (it says a normed field over C is actually equal to C itself)[85], and it turned out to be the first theorem about the general theory of Banach algebras, although one can cite another paper, written in Lvov, on this subject [A: Eidelheit, 1940].

Let us recall that if, as well as taking scalar multiples or performing vector addition, the vectors in a Banach space X can be multiplied in a continuous way, then we arrive at the notion of a Banach algebra. Although the subject of these algebras was considered in Lvov, including S. Mazur's aforementioned presentation and the work of M. Eidelheit, they were only formally introduced in 1941 by the Russian mathematician I. M. Gelfand [B: Shields, 1989]. One of the simplest Banach algebras is the space $C(0,1)$, equipped with the usual multiplication property of functions, while one of the most important ones is the set of all continuous linear operators mapping a Banach space to itself, together with a law of multiplication.

[85] S. Mazur's handwritten copy had a proof of this theorem (for real scalars) but was too long for the requirements of *Comptes Rendus* and the editors asked him to shorten it. Mazur responded by cutting out the proof to make a reduced article, and it was published in that form. The first proof (of the more general theorem, i.e. for complex scalars) was published in [C: Gelfand, 1941], and then the next ones appeared. W. Żelazko recalled Mazur's original proof ([C: Żelazko, 1968; pp. 26–30], [C: Żelazko, 1973; pp. 18–22]). Nowadays this result is called the Mazur-Gelfand theorem.

When A. Turowicz began working at Lvov in 1938, S. Mazur proposed they work together on ring theory. During 1938–1939 they proved a collection of "twenty-something theorems about rings. One of them contained a general theorem that included, as a special case, Weierstrass's Approximation Theorem for continuous functions (arbitrarily many variables)." The paper was completed in 1939, but S. Mazur was very reluctant to publish it. "I don't like to publish straightaway," he reasoned; "something else might spring to mind." A. Turowicz suggested, in 1940, that they publish it in "Studia Mathematica", vol. 9, but Mazur still refused [B: Turowicz, 1995]. While being quite unaware of all this, M. H. Stone proved this important theorem a few years later, and it is now known as the Stone-Weierstrass theorem. In the hope of salvaging some of the other results, A. Turowicz tried to publish them even as late as the 1950s, but he could not manage to persuade S. Mazur even then.

There is a story which still circulates in Polish mathematical circles and of which A. Turowicz was fond: one day in 1938 S. Mazur picked up a new volume of the review journal "Zentralblatt für Mathematik und ihre Grenzgebiete", glanced through the abstracts for papers on convex functions written by French and German mathematicians, and gleefully declared, "Hmm, those results of mine weren't all that bad. They still haven't got everything yet" (cited in [B: Kwapień, 1990; p. 65]).

22.6. Noncommutative probability theory

Geneva hosted an international conference every year during the 1930s, bringing together specialists in a given field. In 1937 the theme of the conference was probability calculus, and H. Steinhaus and W. Heisenberg were among the invited guests. In his talk there W. Heisenberg showed how classical probability theory was incompatible with photon theory. As H. Steinhaus recalls, "Because in my talk I had shown a path that leads to the notion of stochastic independence, but in which the usual laws of probability do not apply, so he was very taken with it" [B: Steinhaus, 2010; p. 150]. That discussion between Heisenberg and Steinhaus, and then the later paper [A: Steinhaus, 1938a] on the subject of independence leading to a noncommutative theory of probability, was the first step in the foundation and development of that theory. Today, the theory plays a major role in applications, particularly in quantum theory.

22.7. Graph theory

As a guest visitor from Lvov, Kuratowski delivered in June 1929 a communication entitled "Sur les courbes gauches" at a meeting of the Warsaw branch of the PTM. In it he presented a theorem stating that a locally connected continuum, containing at most finitely many simple closed curves which cannot be embedded in the plane, must contain one of two figures: either a $K_{3,3}$ graph (formed from the edges of a tetrahedron, together with "crosshandles" joining the midpoints of pairs of opposite edges) or a K_5 graph (formed from the edges of a tetrahedron, together with four line segments linking its center of mass to its vertices). He published the proof a short while later [A: Kuratowski, 1930]. Then he dropped the topic entirely and pursued other interests. Quite possibly he dismissed his result as merely a topological curiosity, but not a particularly interesting one compared with the problem of measure, a problem he was then working on intensively together with S. Banach and S. Ulam (see [A: Banach & Kuratowski, 1929]). In any event, the result remained an isolated case in Kuratowski's mathematical legacy: he never returned to it.

The most widely known version of Kuratowski's theorem is the following (weaker) formulation: a graph is planar if and only if it does not contain a subgraph homeomorphic to either $K_{3,3}$ or to K_5. It is now one of the most basic theorems in modern graph theory, underlined by the fact that it is one of the most frequently cited results in this field (see [C: Berman, 1977]). Other people took Kuratowski's result as a starting point for further investigations, and in time a huge body of literature grew out of it. There were other priority claims, which is nothing new in the history of mathematics, but the conclusions of historical research do tend to rule overwhelmingly in Kuratowski's favor; see [B: Kennedy, Quintas, & Sysło, 1985].

22.8. Applications of mathematics

This was H. Steinhaus's home turf. His intellectual vitality revealed itself in the way he saw mathematics in everything. If it was "mature" mathematics, then he would make it easier to understand and popularize it ([A: Steinhaus, 1938b]), and if he noticed a simple problem, he would look deeper into it and seek a solution. So, for example, when his daughter returned home from school with an assignment to measure the length of the Vistula River from a map, he devised a net for performing the measurement, one which the Książnica-Atlas company later provided with its maps. Working in collaboration with Professor F. Gröer, he took up the problem of finding methods for determining the infection rate of tuberculosis in children. Cooperation with medical doctors prompted him to construct a device for isolating foreign bodies in the body. He built and patented it in 1938, calling it the "introwizor" (he would go on to improve it in Wrocław after the war). He struck upon the idea of working out a new electricity tariff, mathematically simpler and more beneficial for customers who used energy on a regular basis. He invited an electrician called I. Rosenzweig, an assistant at Lvov Polytechnic [A: Rosenzweig & Steinhaus, 1939], to help him physically construct the requisite meters. The idea was to replace the hitherto computationally cumbersome methods, which took irregular energy demands into consideration, with a single, simple formula: the square root of the integral of a quadratic cost function. Already during the war he had formulated an important practical problem, the pragmatic cut: how should one cut a cake (make a good cut) between n people such that everyone would be subjectively satisfied (see [C: Steinhaus, 1948]). The solution for $n = 2$ has been known for ages: one person makes the cut and the other chooses which part to take. He himself found the solution for $n = 3$ during the war, while B. Knaster and S. Banach gave solutions for higher values of n. A complete solution to the problem only appeared half a century later.

H. Steinhaus's credo regarding the application of mathematics was very curious (cit. [B: Urbanik, 1973; p. 49]):

> People go to doctors to seek advice, but not to mathematicians. Would they go to see a doctor if they could only attend medical lectures and if there were no hospitals, clinics, or pharmacies, if the only people available were professors or teachers of hygiene, but nobody who could actually examine them and prescribe them medicine? Until there are practitioner-mathematicians,

> that is to say, trained people capable of providing mathematical advice, and as long as there is no obvious place to seek aid for mathematical ailments in every factory, every office, bureau, bank, or in line management, then all high-brow or low-level didactics will be ineffective.

The wider application of mathematics only developed in Poland after the war, especially in Wrocław (where Steinhaus settled) and partially in Warsaw (where he had students), but that is another story.

22.9. Numerical analysis

The most frequently cited work of S. Kaczmarz is a short note [A: Kaczmarz, 1937], in which he described a method that provides approximate solutions to systems of linear equations in many variables. The method relied on the observation that the problem of determining a solution to a system of linear equations is equivalent to the problem of finding the coefficients of a point lying on the intersection of the hyperplanes given by the system of linear equations. This created interesting possibilities for generalizations and applications.

It is known as *Kaczmarz's method* or *Kaczmarz's algorithm.* Its attractiveness and significance is evidenced by the many times it has been cited in papers (or included in their titles) as well as the translation of the original note into English, which was undertaken in 1993, a full fifty-six years after it was first published. The method has relevance for today's digital age because it is computationally fast. It was without doubt, together with Banachiewicz's Cracovian calculus[86] and the decoding of Enigma[87], the greatest achievement in numerical analysis made by the Polish mathematical school. See [B: Maligranda, 2007].

[86] Tadeusz Banachiewicz (1882–1954) was an astronomer at the Jagiellonian University in Cracow. He invented a new kind of calculus with matrices, which simplified the way the least squares method was applied. His "Cracovian calculus" was popular with geodesists for some time. Mathematicians grumbled about it because multiplication was not associative. Cracovian calculus was pushed aside with the advent of computers.

[87] The decipherment of a coding machine used in Nazi Germany, undertaken by a team of Poznań mathematicians led by Marian Rejewski (1905–1980), would later have the utmost importance for the future allied cause in World War II. See H. Sebag-Montefiore, *Enigma: The Battle for Code*, London, 2000.

CHAPTER 23

On the Periphery

Beyond the mainstream focus of mathematical life at Lvov, there were still some other, more peripheral, subjects which had their own separate existence and which brought some extra color to the whole place.

Logic. Despite there being an excellent tradition of logic in Lvov, stemming from Kazimierz Twardowski's school of philosophy, the subject remained a somewhat peripheral one, without much in common with mathematics and not having the same status as mathematics. One example of the work done is [A: Żyliński, 1925], in which it is shown that disjunction and binegation (i.e. the negation of an alternative) are the only Sheffer functors applicable to word statements. The work had some importance at the time, but now it is merely of historical curiosity.

Where logic met mathematics, there were some important papers written at Lvov. These included [A: Banach & Tarski, 1924], [A: Kuratowski & Tarski, 1931], and [A: Kuratowski, 1931]. But mathematical logic was pursued more intensely in Warsaw, and there were no followers in Lvov.

When the chair of mathematical logic at UJK became vacant in 1930, A. Tarski and L. Chwistek put their names forward. Tarski had already written several good papers by then, but his greatest work still lay ahead of him. Meanwhile, Chwistek was already well known for his simplification of Russell types, specifically for his construction of a "simpler" theory of types contrasting with the original "branched" theory of types. It was undoubtedly the deciding factor why he, not Tarski, got the chair (and was not the result of anti-Semitic resentment). J. Pepis was at Lvov then, and students followed him, from Cracow to Lvov, in the footsteps of the master, including J. Hercberg, W. Hetper, and J. Skarżeński. Pepis worked on classical undecidability theory (he had a joint paper with Chwistek and corresponded with A. Church). Hetper wrote some papers in reports for the Lvov Society of Science on the topic of semiotic arithmetic and the foundations of semiotics, while Hercberg published only one (joint) paper on the foundations of rational semiotics. Skarżeński did not publish.

Classical analysis. Working in isolation outside the mainstream Lvov school was W. Nikliborc. He concentrated on classical problems of mathematical analysis, theoretical mechanics, and hydrodynamics. His first papers developed approximation methods for the solution of differential equations and so-called hyperharmonic functions, namely the real and imaginary parts of analytical functions in two variables, a subject to which he introduced significant simplifications. Inspired by L. Lichtenstein during his visit to Leipzig, he turned to the subject of equilibrium dynamics of rotating fluids. He obtained original results which then allowed him to make progress on the Three Body Problem: by choosing an appropriate local coordinate system he obtained interesting results regarding a system consisting of a sun and two planets. That curious result nonetheless did not become wider known because W. Nikliborc presented his findings in communications to the scholarly meetings of the PTM with the intention of including them later in a monograph of the Three Body Problem, a monograph he ultimately never managed to complete. He had no pupils or successors.

Mathematical cartography. Antoni Łomnicki was a noted authority in this field and was recognized as such on the international forum, as is evident from his role as a consulting expert acting on behalf of the International Committee of Air Navigation. His textbook [A: Łomnicki, 1927] (updated posthumously in [A: Łomnicki, 1956]) is still regarded today as one of the finest ever expositions of mathematical cartography.

Geometry. One of Auerbach's conjectures on convex geometry has been solved only recently, more than seventy years after it was first stated (see [C: Fusco & Pratelli, 2011]). It has direct relevance to a question of Ulam concerning floating bodies.

Perspective in painting. This was the domain of Kazimierz Bartel. He studied Italian paintings in depth, publishing an eminently admirable work [A: Bartel, 1928] still respected to this day.

Actuarial mathematics. Zygmunt W. Birnbaum and Zbigniew Łomnicki worked in this field. Z. W. Birnbaum was statistically oriented, which later helped him to establish himself as a statistician in the United States. Z. Łomnicki worked on the tariffication of accidents.

Part IV. OBLIVION

CHAPTER 24

Ukrainization the Soviet Way (1939–1941)

Lvov was a major academic and cultural center in the Second Polish Republic, rivaling Cracow for second place. It had several large seats of learning (the university, the polytechnic, the Academy of Veterinary Medicine, the Academy of Foreign Trade, the Academy of Agriculture in Dublany, the Higher School of Forestry, the Conservatoire of the Polish Society of Music, the Institute of Plastic Arts), thirty-six high schools, and numerous other schools. As a multilingual and multicultural city, it had roughly 350,000 inhabitants before the war: 63.5% of them declared Polish as their native language, 24% declared Hebrew or Yiddish, 11.3% declared Ruthenian—but 85% of the population spoke Polish on a day-to-day basis. These figures were about the same for the student population; in the last years before the war over 5,000 young people were studying at UJK, more than 60% of them were Poles, Jews made up about 20%, and the Ukrainians roughly 15% [B: Podraza, 1992; p. 43].

The time when it was possible to concentrate on mathematics ended on September 1, 1939. On September 12, during the September Campaign, the Germans reached the outskirts of the city but were repulsed and so their siege began. The Germans handed over control of the besiegement to the Soviets on September 19. On September 22, Lvov surrendered to the Soviets and the city found itself under Soviet control. Shortly afterwards it was swarming with runaways, and the population jumped to 700,000. Among the runaways were some excellent mathematicians from Warsaw (B. Knaster, S. Saks, E. Szpilrajn (Marczewski). "In normal circumstances we would have done a lot with a team like that" [B: Steinhaus, 2002; p. 197]. But circumstances were anything but normal.

Already as early as October 22, 1939, so-called "elections" took place for the People's Assembly of Western Ukraine (S. Mazur was one of those deputized). During its sole session, held sometime between October 26–28, it passed a unanimous(!) resolution in favor of adjoining Western Ukraine to the USSR. On this basis the Supreme Council of the USSR decided on November 1 to adjoin Western Ukraine to the Soviet Union. That is how the new political order was "legitimized".

In reality the Soviets closed down both UJK and the polytechnic in September 1939. That same year, in November, they replaced them with the Ukrainian University, renaming it Ivan Franko University on January 8, 1940, and the Ukrainian Polytechnical Institute. At the university, the theology department was abolished, the medical department was made into a separate college, while the Departments of Humanities and Natural Sciences were each subdivided into two parts, the latter of them becoming the Department of Physics and Mathematics and the Department of Natural Sciences. S. Banach was made dean of the Department of Physics and Mathematics (he was the only Polish dean at the Ukrainian University), and

M. Zaryckij was made the vice dean. The staff list for the six chairs of mathematics read as follows:[88]

Analysis I: Professor S. Banach (Head), Professor L. Chwistek, Professor W. Orlicz, Professor S. Saks, Professor W. Lewickij, assistant J. Hercberg, assistant M. Sperling[89], assistant L. Sternbach, assistant M. Tymoszek, laboratory technician A. Alexiewicz (later a postgraduate student), laboratory technician S. Leja, laboratory technician A. Plamitzer, senior laboratory technician Z. Zahorski (as of January 27, 1940).

Analysis II: Professor H. Steinhaus (Head), Professor H. Auerbach, Reader M. Eidelheit, assistant M. Stark.

Geometry: Professor S. Mazur (Head), Professor B. Knaster, Reader E. Szpilrajn (Marczewski) (crossed off the list on April 15 due to absence at work [actually, he was sitting in prison], put back on it again on June 1, 1940), assistant M. Wojdysławski (later a postgraduate student).

Algebra: Professor E. Żyliński (Head), Reader W. Hetper (crossed off the list because he never turned up), assistant O. Andruszkiv.

Probability Theory: Professor M. Zaryckij (Head), Professor M. Jacob, Reader O. Pepis(transferred to the chair of geometry on April 15, 1940), assistant P. Jurkewycz.

Mechanics: Professor J. Schauder (Head), Professor W. Milianczuk, assistant M. Mosler, assistant A. Raabe.

The Mathematics Institute of the Ukrainian Academy of Sciences in Kiev set up a functional analysis branch in Lvov, where S. Banach was its appointed head; fellow coworkers there included S. Mazur, W. Orlicz, J. Schauder, and H. Steinhaus [B: Mitropolskij & Breus, 1967].

University lectures, which recommenced on January 15, 1940, could be delivered in the Polish language. There even appeared Volume 9 of "Studia Mathematica" (dated 1940), containing virtually all the material either collated before the war or completed during it, but there was also one (excellent) paper written by Soviet mathematicians [C: Krein & Milman, 1940]. As a concession to the new times, every article appeared with an abstract written in the Ukrainian language.

Life was very far from normal, and one of its characteristic features was the Sovietization "national in appearance, socialist in essence". That was mainly why the Ukrainian language became the vessel of ideological content (Marxism-Leninism was taught in Ukrainian, as was the Soviet version of Ukrainian history, etc.), while Polish remained the language of instruction. The Ukrainians of Lvov were still dissatisfied. They desired a radical Ukrainization, similar to the way in which

[88]I thank Professor J. Prytuła of Ivan Frank University in Lviv for providing me this data.

[89]Maurycy Sperling, master's degree at UJK, obtained in 1932.

the Lithuanization of Vilnius had been undertaken[90], and they ceaselessly tried to stamp out use of the Polish language. It was all a bit too inconvenient for Moscow at the time, as testified by a resolution of the Central Committee of the Communist Party of Ukraine (Bolsheviks), dated August 6, 1940, regarding the "unlawful actions of local authorities in Lvov", and by the decision of the Political Bureau of the All-Union Communist Party (Bolsheviks), on September 8, 1940, to permit lectures to be delivered in Polish, while at the same time encouraging everyone to learn Ukrainian and Russian and to start adopting them when giving lectures [B: Kalbarczyk, 2001]. On March 5, 1941, the new rector (appointed September 23, 1940), Heorhij Stepanovich Bychenko warned the Polish professors they would lose their jobs if they did not start speaking in Ukrainian [B: Draus, 2007].

"The First Academic Session of the Ivan Franko Public University of Lvov" was supposed to show that normal academic life had resumed. It was held between January 29–February 3, 1941. The opening and summary speeches were made by the rector H. Bychenko. Of the five plenary speeches, one was mathematical: Professor M. Zaryckij, *Leninskij analiz osnow statistiki* (Leninist analysis of the foundations of statistics). The talks were organized in sessions. Professor S. Banach was in charge of the mathematical talks; the program is given below (titles have been translated into English):

Professor S. Mazur, *On fundamental notions in the geometry of differentiable curves.*

Professor B. Knaster, *On Antoine's arc.*

Reader J. Pepis, *On recursive functions and recursive equations.*

Reader E. Szpilrajn (Marczewski), *On unsolved problems in measure theory.*

Assistant M. Mosler, *On a theorem of Bernstein.*

Professor M. Jacob, *The Gibbs phenomenon for orthogonal polynomials.*

Professor E. Żyliński, *On equations of degree three.*

[90] Vilnius, like Lvov, has a complicated history. It was the capital of the Grand Duchy of Lithuania, which was joined with the Kingdom of Poland to form a single state, called the Republic of Two Nations (Rzeczpospolita Obojga Narodów). After a few generations the Lithuanian elite had become less influential and had lost out to the Polish presence. When the University of Vilnius was established in 1578, the language of instruction was Latin but was later changed to Polish. Vilnius became a Polish city. After World War I, Vilnius became part of Poland, but Lithuania never accepted this. In September 1939 the Soviets took over Vilnius but then gave it to Lithuania on October 10, 1939. The Lithuanians then immediately began eradicating all traces of Polishness. On December 15, 1939, they closed down Stefan Batory University (denoted SBU), dismissed the staff, and expelled the students (numbering about 3,000). They replaced SBU with a Lithuanian university, and although they re-employed some Polish academics and re-admitted some of the Polish students (141 of them, according to the records for February 1940), it became a completely different seat of learning: Lithuanian and anti-Polish. In June 1940 the Soviets seized all of Lithuania, together with Latvia and Estonia, and annexed all these states to the USSR.

Professor H. Steinhaus, *On the density distributions that arise from projections of uniform distributions.*

Professor I. (sic!) Auerbach, *On curves and surfaces having conjugate diameters.*

Professor S. Banach, *On Baire functions in linear spaces.*

Professor L. Chwistek, *On the foundations of quantum mechanics.*

Professor W. Orlicz, *On Hölder functions.*

Professor S. Saks, *On subharmonic functions.*

Assistant Z. Zahorski, *On the zero set of derivatives.*

Assistant. J. Hercberg, *On the fundamentals of algebraic analysis.*

The talks were held over three days (January 30–31 and February 1) between 2:00–4:00 pm and 7:00–9:00 pm (Moscow time). One hour was allotted for each talk. As one might imagine, judging by the titles, most of them were overviews. Everything went without incident, and later nobody thought about them again.

Academic visits were relatively frequent. Among those who visited Lvov were N. N. Bogolubov, P. S. Aleksandrov, S. Sobolev, A. F. Bermant, L. Lusternik, and M. Krein. From Lvov, E. Żyliński traveled to Moscow on a visit (November–December 1940), as did S. Banach, S. Mazur, and J. Schauder (January–February 1941). S. Banach, S. Mazur, and M. Zaryckij visited Kiev (they went there together in December 1940). S. Banach and M. Zaryckij visited Tbilisi (March 1941), after which S. Banach went on alone on a brief visit to Odessa. Banach was back again in Kiev on June 19, 1941, but immediately returned home to Lvov upon hearing news of the German invasion. Attending the Ukrainian University at the time were students and postgraduate students who would later become well-known mathematicians and physicists, including J. Albrycht, A. Alexiewicz, S. Hartman (year three), R. S. Ingarden, M. Nosarzewska (year four), M. Wojdysławski, Z. Zahorski (all later became known in Poland), and W. Lance, M. Wiszyk, A. Strauss (M. Wiszyk and A. Strauss became known in the Soviet Union).

Life was a total misery. There were endless meetings and reschedulings. There was the perpetual threat of prison, deportation, or arrest without warning. Deportations (which were undertaken in four mass waves: February 8–9, 1940; April 12–15, 1940; June 20–30, 1940; and June 1941) and mass migrations caused the population of Lvov to drop to 400,000 in 1941. The principal victims of Soviet oppression were the Poles. For example, the students among them at the university amounted to just 362 ([B: Olesiak, 2001], [B: Podraza, 1992; p. 43]). One should really add to that number some of the Poles who belonged to the 1,400 or so students at the Medical Institute. In any case, the fall in the number of Polish students was striking: comparing the records for 1939 and 1941, there was almost a ten-fold drop, from over 3,500 to just under 400. The student population in general fell as well: in June 1941 there were only 1,617 students attending Lvov University, 44% of whom were Jews, 34% Ukrainians, and only 22% Poles.

> I acquired an unshakeable, truly physical revulsion to all the Soviet bureaucrats, politicians, and commissars. I saw them as slow-witted, deceitful, stupid barbarians who had us in their grasp, like the giant ape which kidnapped Gulliver and whisked him up onto a rooftop.
>
> [B: Steinhaus, 2002: p. 191]

K. Bartel spent time in a Soviet prison (December 1939), as did E. Szpilrajn (Marczewski) (spring 1940, after trying to get to Hungary), and the first fatalities occurred.

The following Lvov mathematicians did not survive the Soviet occupation (1939–1941):

Władysław HETPER (arrested, died in a Gulag, 1940),

Stefan KACZMARZ (perished in 1939, under unknown circumstances).

CHAPTER 25

The German Occupation (1941–1944)

The German occupation of Lvov started one week after the start of the Russo-German War and lasted more than three years, specifically from June 30, 1941, to July 27, 1944. It was the second stage of the eradication of Poles and Polish culture in that city.

The first liquidations of people took place as early as July 1941, when the Germans, aided by Ukrainians, murdered more than three thousand Poles and Jews[91], including over 200 UJK students[92]. In the shadow of this mass murder, twenty-three professors from the university, the polytechnic, the veterinary academy, and the Academy of International Trade were arrested on the basis of a list that had been drawn up beforehand. Almost all of them were shot[93] immediately afterwards.

One of the first to be arrested, as early as July 2, was Professor Bartel, head of the descriptive geometry section at the polytechnic. He had earlier served as a government minister, a senator, and had been the prime minister of the Republic of Poland three times. Yet he was one of the last of the group to be shot: he was executed on July 26. He was killed later than the others because the Germans offered to spare his life in exchange for his collaboration, an offer he declined.

On the night of July 3 several dozen professors were arrested. On the morning of July 4 all of them (with one exception) were shot. Included among them were the following mathematicians: Professor Antoni Marian Łomnicki and Professor Włodzimierz Stożek, two of the directors of mathematical chairs at the polytechnic (Stożek's sons died with him: Eustachy, aged 29, was an assistant at the polytechnic, and Emanuel, aged 24, was a PL graduate). Professor Franciszek Groër was the only person to avoid the same fate as his colleagues, due to his German ancestry.

Professor A. Łomnicki's wife recalls (quoted from [B: Kozielecki, 1999; p. 35]; see [B: Maligranda, 2007]):

> Sleep was impossible. I stood long hours trembling by the window, waiting for the day when I could get out and try to make sense of this invasion. When dawn broke I saw, from the window of my flat on the third floor, signs of movement on the Wulecki Heights. They turned out to be silhouettes of people [...]. I sat down on the sofa, wondering what this movement could be for at such an early hour, because it was about four o'clock in the morning. At that split second I heard the first rifle salvo—I

[91] *The PWN Great Encyclopedia*, Vol. 16, p. 243 [in Polish].

[92] Hundreds of UJK students were arrested on July 11, 1941. They were first taken to the Ukrainian military police station on Łoziński Street and then transferred to the Gestapo prison (on Pełczyńska Street). Many did not return, and according to Polish Underground sources over two hundred of them were shot.

[93] [B: *Leopolis docet*, 2001].

FIGURE 23. Monument, in Wrocław, in memory of the professors murdered in Lvov. Sculpted by Borys Michałowicz, erected (1964) during Poland's communist period, between two buildings of Wrocław University of Technology, by Grunwald Square [B: Mierzecki, 2009]. Because the authorities at the time did not permit any reference to Lvov to appear on the momument, all that did appear on the plaque were the enigmatic words "Our fate—a warning" ["Nasz los—przestroga"], understood only by those who knew the real truth. Acknowledgment of the execution, including the names of the individuals executed, only appeared beside the monument in 1981.

> knew immediately what it meant—I ran out onto the staircase, which had windows peering out slightly further to the right, giving me a wider view. And then I saw that the people who were coming down from the top were stopping somewhere about halfway down, in a small hollow. I recognized German soldiers, next came men in civilian clothes, there were some women as well, and I was particularly struck by the figure of a priest wearing a soutanne. I was later told that that was Father Komornicki [...]. I saw a figure in grey clothes, the total likeness of my husband, but I didn't want to let that thought enter my mind. Several times people were presented five at a time and I saw how those people fell with each rifle salvo. I stood rooted to the spot, not consciously looking at the scene of execution, while with me there were two other ladies from the neighborhood.

Professor Stanisław Ruziewicz was arrested (one of two) on July 11. He was the head of one of the UJK chairs of mathematics during 1921–1934, subsequently

FIGURE 24a. Wulecki Heights

head of the chair of applied mathematics at the Academy of International Trade, becoming its rector in 1939. He was shot the very next day after his arrest.

The circumstances of this atrocity remain unclear to this day, though the collaboration of Ukrainian nationalists does now appear to be beyond all reasonable doubt. According to S. S. Nicieja, the list was "drawn up by Ukrainian nationalists biding their time at the OUN (Organization of Ukrainian Nationalists) center in Cracow and by former students of Lvov colleges who had escaped Lvov before it was taken over by the Red Army. Here the vested interests of the Nazis and of OUN members coincided. The Germans wanted to liquidate the Polish intelligentsia;

FIGURE 24b. Plaque at the Wulecki Heights.

the Ukrainians sought the de-Polonification of Lvov" [B: Nicieja, 2001]. The people who compiled the list remain unknown to this day. Ukrainian collaboration is in any case indicated by the Ukrainians themselves; for example Boris Lewickyj states, "Ukrainians from nationalist circles helped the Germans by drawing up a list of members of the Polish intelligentsia, the de-Polonification of Lvov being their motivation" [B: Lewickyj, 1960]. The manner by which the de-Polonification was carried out indicates that there were cases in which adult family members, guests, and attendants found at the flats of arrested professors were also taken away and

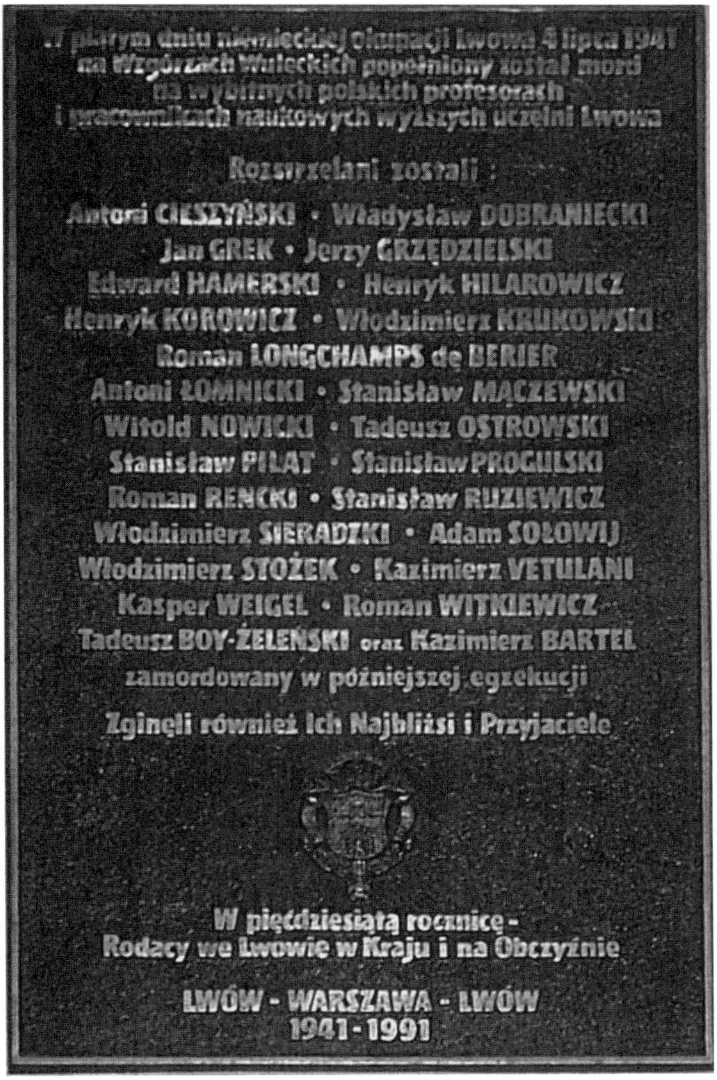

FIGURE 25. Memorial plaque in Lvov cathedral. The Soviet authorities, and for the first few years the Ukrainian authorities also, refused to give permission for a monument in memory of the execution of "Polish professors". In response to that situation, a suitable plaque was mounted in Lvov's Roman Catholic cathedral. However, that plaque contains omissions and inaccuracies.

subsequently murdered. It was the first sign of "contemptible behavior by parts of the Ukrainian intelligentsia during the period of occupation," regarding which B. Lewickyj spared neither words nor criticism[94]. According to K. Lanckorońska[95], there was even a second list of professors, but it never got used.

---

[94]See B. Łewickyj's letter to J. Giedrioyc, dated November–December 1965. I quote from B. Łewickyj, *Terror*, ..., p. 14.

[95][B: Lanckorońska, 2008].

FIGURE 26. July 3, 2011: the seventieth anniversary of the murder of the Lvov professors is marked by the unveiling of a new monument on the site of the execution at the Wulecki Heights. Attendees included prominent Polish and Ukrainian officials. Cutting the ceremonial ribbon were Rafał Dutkiewicz, Mayor of Wrocław, and Andriy Sadovyi, Mayor of Lviv.

In Wrocław there now stands a monument (see Figure 23) dedicated to the memory of the Lvov professors and their family members who were executed[96] A memorial plaque hangs in the Collegium Maius at the Jagiellonian University (UJ) in Cracow, as does another in Lvov cathedral (Figure 25). A modest memorial stands on the Wulecki Heights, the actual site of the murders (Figure 24a). A new monument was unveiled there on July 3, 2011 (see Figure 26).

On July 17, 1941, East Galicia became subject to the General Governorship [of the Third Reich]. There then followed a systematic extermination of the Jewish population, and part of the Polish one also. It is estimated that some 160,000 inhabitants of Lvov perished during 1941–1944[97].

H. Steinhaus quickly sensed the danger. He wasted no time in burning all personal documents and then on July 4, 1941, he fled, together with his wife, his flat at 14 Kadecki Street, never to return again. The two of them hid for a few days with friends; then they hid at Professor B. Fuliński's place until November 1941. When that became too risky, they moved first to Osiczyna, near Żimna Woda, and then in July 1942 they went to Berdechów, near Stroże. Hugo Steinhaus assumed the name Grzegorz Krochmalny throughout the period 1941–1945. They both survived the war but never returned to Lvov after its conclusion (they settled in Wrocław, where Steinhaus became the first dean of the Department of Mathematics, Physics, and Chemistry at the new Polish colleges, the university, and the polytechnic).

---

[96]See [B: Mierzecki, 2009].

[97]The *PWN Great Encyclopadia* [Polish], Volume 16, pp. 243–244.

The Lvov colleges, including the university and the polytechnic, were closed down by the Germans (only elementary schools continued because no postelementary teaching was planned for locals), but in spring 1942 a two-year State Commercial School in Polish [Staatliche Handelsschule mit Polnischer Unterrichtssprache] was started, in which W. Orlicz taught courses in accountancy for tradesmen, as well as the State Professional Courses [Staatliche Fachkurse], for which four-year courses (in five fields) were organized at the polytechnic, in medicine, veterinary science, and forestry. The programs were taken from prewar schools of higher learning, but those attending them had no right to switch to German colleges. Seven hundred twenty-nine people registered for those courses. W. Nikliborc taught mathematics, and M. Zaryckij followed after him.

A singular feature of the German occupation was the institute of Professor R. Weigl, who produced antityphus vaccines for the needs of the Wehrmacht. The Weigl Institute employed, as louse feeders, outstanding representatives of the Lvov intelligentsia. The mathematicians among them included S. Banach (over the period 1942/1943; then Banach fell ill), B. Knaster, and W. Orlicz, and from the younger ones there were J. Albrycht and F. Barański (according to the recollections of Professor W. Szybalski, who was the "head of the lice lab" during that period)[98].

Even in July 1941, Edmund Bulanda, the predecessor of the last UJK rector, Roman Longchamps de Bérier, who was executed on the Wulecki Heights, tried to reactivate the UJK on a clandestine basis. He persuaded professors from each of the five prewar UJK departments to work with him, and the underground university started up. August Zierhoffer, a geographer (in this capacity in 1939) was appointed dean of the mathematics and natural sciences department. Mathematics lectures were delivered by, among others, Eustachy Żyliński and Władysław Orlicz [B: Draus, 2007].

The following Lvov mathematicians did not survive the German occupation (1941–1944):

Hermann AUERBACH (poisoned himself in 1942 while awaiting imminent deportation to the extermination camps)[99],

Kazimierz BARTEL (executed by firing squad, 1941),

Meier EIDELHEIT (murdered by the Germans, 1943),

Jan HERCBERG (missing/presumed dead, 1941),

Marian Mojżesz JACOB (missing/presumed dead, 1944),

Antoni ŁOMNICKI (executed by firing squad, 1941),

Stanisław RUZIEWICZ (executed by firing squad, 1941),

Juliusz SCHAUDER (murdered by the Germans, 1943),

Józef SCHREIER (suicide by poisoning, 1943),

Ludwik STERNBACH (missing/presumed dead, 1942),

---

[98]Details of the work done at the Weigl Institute are described in the personal recollections [B: Krawczyński, 2009] of someone who was gradually promoted at the institute, starting as a stoker and working his way up to the position of assistant to the director responsible for the louse-feeding section.

[99]In 1945 S. Banach gave evidence before a Soviet prosecutor, testifying that H. Auerbach had been lying in Rapaport Hospital when he was shot "in action", together with all the hospital staff and patients (Professor J. Prytuła has a copy of these proceedings).

Włodzimierz STOŻEK (executed [with his two sons] by firing squad, 1941),
Kaspar WEIGL (executed [with his son] by firing squad, 1941),
Menachem WOJDYSŁAWSKI (missing/presumed dead, 1942).

CHAPTER 26

The Expulsion of Poles (1945–1946)

Already in 1943, the Big Three decreed in Tehran that "in principle, [...] the seat of the Polish nation state should be located somewhere between the so-called Curzon Line and the river Odra"[100].

In July 1944 the population of Lvov dropped below 150,000 people. The Soviet Army reached the outskirts of the city on July 22, 1944, and captured it after five days' fighting, with substantial support from the Polish Home Army. For example, the Kedyw Company (of the Polish Home Army) captured the main building of the polytechnic and subsequently set up a Home Army military hospital there. As early as July 31, the Soviets nonetheless arrested all officers in the Polish Home Army, and a new order of life had begun. It was the Soviet way of doing things, a characteristic feature being the outward semblance of normality matched with a ruthless realization of long-term political goals.

The colleges started functioning again—the university opened in September, the polytechnic in October 1944, followed by the Academy of Veterinary Medicine and the Academy of Foreign Trade. Despite the war losses, Poles still made up the largest number of lecturers at these colleges[101] as well as the largest number (almost 80%) of students[102].

The Department of Mathematics and Mechanics at the Ukrainian University of Lvov for the academic year 1944/1945 was as follows:

Mathematical Analysis: Professor S. Banach (Head), Zyrianowa (assistant), A. Alexiewicz (senior laboratory technician).

Theory of Functions: Acting Professor W. Orlicz (Head), Openheim-Dawydowa (laboratory technician).

Geometry: Professor S. Mazur (Head), Acting Professor B. Knaster, Jasienicki (assistant).

Algebra: Acting Professor Żyliński (Head), Towarnickij (assistant).

Theoretical Mechanics: Professor W. Nikliborc (Acting Head).

General Mathematics: Acting Professor M. Zaryckij (Head), Ripun (assistant), Trochimczuk (laboratory technician).

[100]See [B: Kalbarczyk, 2001; p. 212]

[101]According to the records for April 1, 1945, there were at that time 819 lecturers at Lvov colleges, of whom 391 were Poles and 306 were Ukrainians [B: Kalbarczyk, 2001; p. 214].

[102]See [B: Olesiak, 2001].

The faculty heads were generally Poles (including five heads of six sections), but their situation quickly took a turn for the worse. Increasingly confident about the territories they had been acquiring since 1939, the Soviets acted on the basis of done deals. As early as July 26, 1944, they made a (secret!) border agreement with the PKWN (Polish National Liberation Committee, a puppet government installed by the Soviets in July 1944). A month later they made another deal, regarding the resettlement of the population. Meanwhile, war was still being waged; for almost half a year the war front stayed stuck between the San and Vistula Rivers. At the same time, in Lvov, the Soviets were preparing to resettle the Poles elsewhere for good: in theory there was a "free choice" to stay or go to the "new Poland". Lectures were now ruthlessly required to be delivered in the Russian or Ukrainian language. Those who refused to cooperate were deemed unfit to fulfil their statutory obligations and were therefore removed from their colleges. Professor Stanisław Mazur was summoned to Moscow in October 1944, where he was informed about the new Polish borders and about the plans to resettle Poles. On December 6, 1944, at a local meeting of the Polish intelligentsia, the secretary of the Lvov District Committee KPU(b), Ivan Hruszewski, dispelled any lingering doubts: it was a case of leave or be Sovietized. Among other things, he told them, "The Polish population should understand that Lvov was, is, and always will be a Soviet city. That being the case, it will be run by Soviets [...]. There will be no more nonsense, as in the years 1939–1941 [...]. In accordance with the agreement made with the PKWN, we are giving every honest Pole the freedom to choose which state to belong to" [B: Kalbarczyk, 2001, pp. 215–216]. But the Poles had little desire to exercise their "freedom of choice" and abandon their home city. It was on the night of July 3–4, 1945, when a wide-scale operation of mass arrests, targeting so-called "Polish nationalists", was enacted. Over 2,000 people were rounded up, including twenty-five academics. Over 100 scholars signed a petition of protest to Bolesław Bierut, the head of the KRN (Nationalist County Council, a puppet communist parliament). Professor Stanisław Mazur also intervened, which may be why twenty-two of the twenty-five-strong group of academics had been released by the end of 1946 (one died in the camps, but it is not known what happened to the remaining two). However, a worse fate awaited the rest; some returned only after many years, and others never returned at all. The Lvov newspapers of February 13, 1945, reported on the Yalta Conference, confirming the decisions that had been made at Tehran. Shortly afterwards the Polish inhabitants of Lvov began to leave, including the academics.

Once Lvov had been freed from the Germans, the few Polish mathematicians still left alive returned to academic work. S. Banach began giving seminars. Between September 25–October 2, 1944, S. Banach and S. Mazur took part in the twenty-fifth anniversary of the Ukrainian Academy of Sciences in Kiev; at the beginning of 1945 (in January–February) Banach and his wife were in Moscow. However, Banach's illness was apparent by then. His condition continued to deteriorate, and on November 17, 1944, he requested that he be relinquished from his official duties as dean (M. Zaryckij became the new dean); then he resigned from all his other duties (he died on August 31, 1945). B. Knaster obtained a professorship on January 27, 1945, but on April 4, 1945, he gave it up, as he was going to leave. Earlier, on February 10, Orlicz had tendered his resignation, since he too was leaving. S. Mazur and E. Żyliński left on May 1, 1945, but returned to their

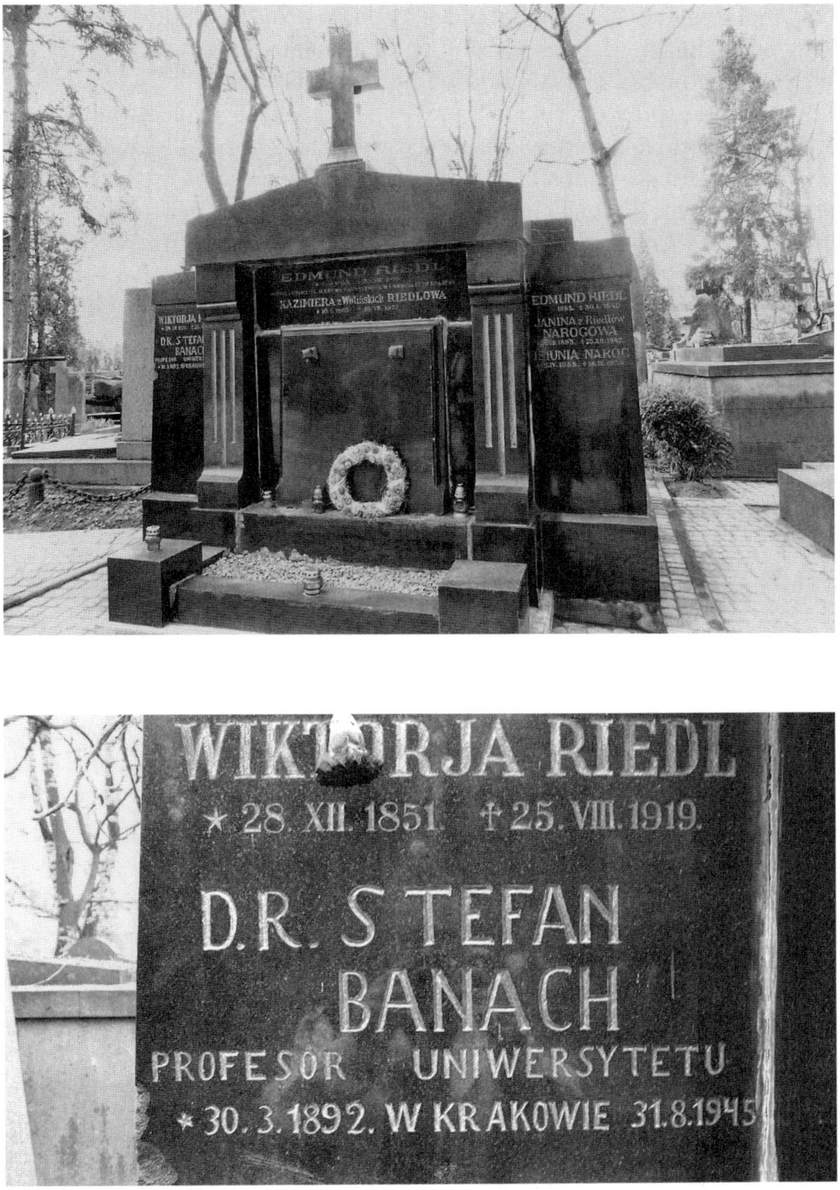

FIGURE 27. Stefan Banach's grave in the Łyczakowski Cemetery.

jobs and postponed their departures (Mazur left in May 1946). The vacant departmental chairs were filled: replacing Banach—W. Lewickij; replacing W. Orlicz—A. S. Kowańko; replacing S. Mazur—W. Bujmola, though S. M. Nikolski[103] had

[103]Siergej Michailovitch Nikolski (1905–2012). In 1929 he completed studies at a school of higher education (later it became a university) in Dniepropietrovsk. He worked there until 1940 and then at the Steklov Institute of Mathematics. He was only at Lvov for a short time. He was a member of the Russian Academy of Sciences, the Hungarian Academy of Sciences, and the Polish Academy of Sciences. His principal contributions were in functional analysis, the theory

been the replacement originally intended; replacing W. Nikliborc—G. M. Sawin[104]; replacing E. Żyliński—J. Łopatyńskij[105]. The chair of probability and statistics restarted on September 1, 1945, headed by B. W. Gnedenko[106].

The first transport of "repatriated" Polish academics rolled out on April 18, 1945. The next one followed sometime between the end of May and the beginning of June. But the real exodus began after the announcement of the Declaration of Potsdam in August 1945. The last transport left Lvov in February 1946. Only a handful of Lvov's Polish scholars remained. As Professor J. Rudawśkij explained, after the Poles had been thrown out, Lvov barely had 30,000 inhabitants left[107].

---

of approximation of one-variable and many-variable functions, partial differential equations, and variational calculus.

[104]Gurij Mikolajevich Sawin (1907–1975). He completed mechanics studies in 1932 at the Institute of Mathematics, Physics, and Chemistry in Dniepropietrovsk. He worked there until 1941, at the Institute of Engineering and Construction. He obtained his doctoral degree (equivalent to a habilitation degree) in 1940 at the Mathematics Institute of the Georgian Academy of Sciences, in Tbilisi. He was the rector of Lvov University during 1949–1952, and between 1945–1949 he held the chair of mechanics there. His later years were spent in Kiev, where he worked at the university and at the Ukrainian Academy of Sciences, having been a correspondent member of it since 1948. He authored over 300 works, including several books.

[105]Jurij Jarosław Łopatyńskij (1906–1981). He completed studies in mathematics and physics at the university in Baku in 1926, subsequently working there until 1946. He obtained the status of science candidate in 1938 (at Moscow University), and the title of professor in 1947. During 1946–1963 he worked at the university in Lvov. He was a member of the Ukrainian Academy of Sciences in 1951 (correspondent member in 1951, full member in 1965). He authored sixty works and several monographs.

[106]Boris Vladimirovich Gnedenko (1911–1995). He completed studies at the University of Saratov in 1930 and worked at Moscow University between 1934–1945 and from 1960 onwards (he became a professor there in 1942). He was at Lvov University during 1945–1950 and at Kiev University during 1950–1958. He was a member of the Ukrainian Academy of Sciences. His principal works are in probability theory.

[107]Talk given by the rector of Lviv Polytechnic, Professor Jurij Rudawśkij, at a conference in Wrocław in 2001.

Part V. HISTORICAL SIGNIFICANCE

CHAPTER 27

Chronological Overview

This chronological overview considers the years 1919–1945. Key information about Lvov mathematicians, including doctorates, habilitations, professorships as well as some publications and results are given in their chronological order.

1919

Józef Puzyna dies. After W. Sierpiński leaves for Warsaw and after the death of J. Puzyna, the two chairs of mathematics at Lvov university remain vacant. Eustachy Żyliński is appointed to the first chair, as an ordinary professor, which fills the gap to a certain extent.

Antoni Łomnicki obtains his habilitation at Lvov Polytechnic.

The first paper by a Polish author on functional analysis appears [A: Steinhaus, 1919].

S. Banach writes his first paper [A: Banach & Steinhaus, 1918] (it came out in 1919). It is a joint paper with H. Steinhaus and was written as early as 1916 [B: Steinhaus, 1961a; p. 252]. Formally speaking, S. Banach was still a student at the time of its printing.

1920

Zygmunt Janiszewski dies. Hugo Steinhaus is asked to become professor extraordinary at UJK University, i.e. Jan Kazimierz University in Lvov (as it would be called from that time onwards until 1939), and A. Łomnicki is appointed professor extraordinary at PL (Lvov Politechnic, known up to then as the Polytechnic School).

Stefan Banach obtains his doctorate, and Steinhaus arranged a position for him as assistant to Professor Łomnicki. Steinhaus's paper on the set of distances of points in a set of positive measure appears [A: Steinhaus, 1920; Th. VIII].

1921

Antoni Plamitzer obtains his habilitation (in descriptive geometry) at PL.

Antoni Łomnicki becomes an ordinary professor at PL.

The Lvov Mathematical Association, founded in 1917, becomes the Lvov branch of the Polish Mathematical Society.

1922

Eustachy Żylinski becomes an ordinary professor at UJK. Stefan Banach obtains his habilitation and becomes a professor extraordinary at UJK.

Włodzimierz Stożek from Cracow and Antoni Plamitzer are appointed ordinary professors at PL.

Stefan Banach's doctoral thesis is published [A: Banach, 1922a], the doctorate having been awarded at UJK in 1920. It is his seventh contribution to mathematics, but his first work of world stature. It marks the start of the theory of Banach spaces (see Chapter 14).

1923

Hugo Steinhaus becomes an ordinary professor at UJK. Adam Maksymowicz obtains his habilitation at PL.

Publication of the papers [A: Łomnicki, 1923] and [A: Steinhaus, 1923a], containing a measure-theoretic approach to probability calculus which later becomes more commonly adopted (see Chapter 16).

Stefan Banach investigates the general problem of measure theory. He completes the remaining unsolved cases of a problem addressed in Hausdorff's 1914 paper: whereas Hausdorff had shown the problem has an affirmative answer for the case $n \geq 3$, Banach now proves there is a negative answer for the cases $n = 1$ and $n = 2$ [A: Banach, 1923] (see Chapter 16).

1924

Stanisław Ruziewicz is appointed as an ordinary professor at UJK.

Stefan Banach obtains a year's academic leave to continue his research in Paris.

Representatives of the younger generation obtain their doctorates: Stefan Kaczmarz, Władysław Nikliborc, and Juliusz Schauder.

The paper [A: Banach & Tarski, 1924] contains an astonishing result, known ever since as the Banach-Tarski paradox (see Chapter 16).

1925

A short note [A: Steinhaus, 1925] appears, containing pioneering ideas about the then nonexistent theory of games (see Chapter 17).

1926

Włodzimierz Stożek is appointed as an ordinary professor at PL.

J. Schauder's doctoral thesis, awarded in 1924 at UJK, is published [A: Schauder, 1926].

1927

Stefan Banach is appointed as an ordinary professor at UJK. At the same time, four chairs in mathematics are established at UJK and are assumed by Stefan Banach, Stanisław Ruziewicz, Hugo Steinhaus, Eustachy Żylinski. This arrangement lasts until 1933.

Antoni Plamitzer becomes the third ordinary professor, joining Antoni Łomnicki and Władysław Stożek, at PL. At the same time Kazimierz Kuratowski becomes an ordinary professor (the fourth one) at PL and assumes responsibility for the mathematics chair at the Department of General Studies, a role he would fulfil until 1933.

Juliusz Schauder obtains his habilitation. Although a great talent, he was only to give lectures on a commission basis at UJK, while earning his daily bread as a school teacher. Władysław Nikliborc obtains his habilitation in mathematics at PL (in 1931 he would also obtain his habilitation from there in theoretical mechanics).

The First Congress of Polish Mathematics takes place in Lvov, with the participation of some 200 individuals, including among them some renowned mathematicians from abroad, such as John von Neumann, Nikolai Łusin, Nina Bari, Vaclav Hlavaty [B: Księga Pamiątkowa, 1929].

S. Banach and H. Steinhaus publish their excellent "theorem on the condensation of singularities" [A: Banach & Steinhaus, 1927] (see Chapter 18).

The paper [A: Schauder, 1927] appears, translating Brouwer's fixed point theorem to Banach spaces. It is the beginning of the use of topological methods for such spaces and also the beginning of nonlinear functional analysis.

1928

Władysław Orlicz obtains his doctorate and leaves for Göttingen for a period of two years.

A sizeable contingent of mathematicians from Lvov travels to the International Congress of Mathematicians in Bologna: S. Banach, K. Bartel, L. Chwistek, S. Kaczmarz, A. Łomnicki, Z. Łomnicki, W. Nikliborc, S. Ruziewicz, H. Steinhaus, W. Stożek, E. Żyliński. Kaczmarz and Steinhaus deliver sessional talks, which are based on concepts formulated by Banach.

In 1911, O. Toeplitz commenced his general theory of summability. In 1926, H. Steinhaus, who knew Toeplitz and his ideas, gave S. Mazur a certain problem relating to them. Mazur now publishes his first paper on this subject [A: Mazur, 1928].

1929

Zygmunt Wilhelm Birnbaum obtains his doctorate.

Stefan Kaczmarz obtains his habilitation.

At the First Congress of Mathematicians from Slavonic Countries, held in Warsaw, those in attendance from Lvov were S. Banach, S. Kaczmarz, K. Kuratowski, A. Łomnicki, W. Nikliborc, H. Steinhaus, W. Stożek.

The first volume of "Studia Mathematica" appears. It contains fourteen papers, twelve of which were from Lvov (Banach, two; Birnbaum, one; Kaczmarz, one; Mazur, one; Nikliborc, two; Schauder, one; Sternbach, two).

The most important contribution to that volume is undoubtedly Banach's paper [A: Banach, 1929a], in which Banach considers linear and continuous operators (on Banach spaces) and proves many theorems, including the one later known as the Hahn-Banach theorem on the extension of functionals. It introduces the very important notion of a dual space and dual operator (see Chapter 18), thereby initiating duality theory.

H. Steinhaus proposes that the "type B spaces" (in Banach's terminology) be called "Banach spaces" [A: Steinhaus, 1929]. The name is quickly adopted and is commonly used to this day.

Mazur's paper [A: Mazur, 1929] initiates research into the topological structure of linear metric spaces and their subsets (see Chapter 18), which later blossoms into infinite-dimensional topology.

Banach reformulates the Hausdorff measure problem in terms of a more set-theoretical one, and in a joint paper [A: Banach & Kuratowski, 1929] the authors show that if the continuum hypothesis is true, then sets having the power of the continuum do not have a measure, in the sense of Banach, not even a measure that varies under translations.

A short paper appears with a new proof of the fixed point theorem [A: Knaster & Kuratowski & Mazurkiewicz, 1929]. It contained highly influential ideas on fixed point theory and related problems [B: Granas, 1981]. Although at that time only one of its three authors had a permanent link with Lvov, the type of problems considered in that paper were being investigated just as much in Warsaw as in Lvov.

1930

Herman Auerbach and Miron Zarzycki (Zaryckij) obtain their doctorates.

Kazmierz Bartel is elected president of the Polish Mathematical Society, his term of office running between 1930–1932.

Stefan Banach is awarded the Benedykt Dybowski prize by the city of Lvov.

Leon Chwistek is called to serve as the chair of logic at UJK (as professor extraordinary). Following in Chwistek's wake are his students Jan Hercberg, Władysław Hetper, and Jan Skarżenski, young mathematicians interested in logic.

M. Zarzycki participates in the all-union mathematical congress in Charkov. Ulam's paper appears [A: Ulam, 1930], containing the most important results of his doctoral thesis [A: Ulam, 1977], which he obtained in 1933 after passing all his requisite exams.

The second volume of "Studia Mathematica" appears. It contains twenty-four papers, of which eighteen come from Lvov (Auerbach, one; Banach, three; Banach and Saks, one; Birnbaum and Orlicz, one; Kaczmarz, one; Kaczmarz and Steinhaus, one; Mazur, four; Orlicz, two; Schauder, three; Schreier, one).

Schauder's papers deliver important complements to Banach's results. Schauder simplifies Banach's theorem on the invertibility of linear and bijective operators and shows that it leads to an important theorem on domain invariance [A: Schauder, 1930a]. In the second paper he generalizes Banach's contraction theorem [A: Schauder, 1930b] (see Chapter 18).

S. Mazur introduces convexity arguments into functional analysis [A: Mazur, 1930a]. In another paper he comes up with the wonderful idea of applying Banach space theory to summability theory [A: Mazur, 1930b]. The realization of his idea (by S. Mazur himself, W. Orlicz, and others) ensures that summability theory becomes subsumed as part of functional analysis (see Chapter 18).

S. Banach generalizes the measure problem even further, in a more set-theoretic setting [A: Banach, 1930a]. It is followed up in [C: Tarski, 1930] and [A: Ulam, 1930], and problems in measurable cardinal numbers and real-valued measurable cardinal numbers, numbers which arose as a result of such work, became part of the foundations of mathematics. But this was not to be of further interest to Banach.

A. Zygmund starts a close working relationship with the Lvov center, publishing a monumental work on orthogonal systems [C: Zygmunt, 1930].

1931

S. Banach, M. Jacob, and H. Steinhaus travel from Lvov to attend the Second Congress of Polish Mathematics, held in Vilnius [B: Drugi Zjazd, 1931].

Volume 3 of "Studia Mathematica" appears. It contains seventeen papers, eight of which come from Lvov (Auerbach and Banach, one; Banach, two; Birnbaum and Orlicz, one; Kaczmarz, two; Orlicz, one; Ruziewicz, one).

Some of these papers, including [A: Banach, 1931b], apply the Baire category to function spaces. The Polish School of Mathematics (equally in Warsaw as in Lvov) fashioned the Baire category into an eminently useful tool, particularly for proving existence theorems. As a result, the spaces which brought Polish mathematicians their greatest successess, such as metric, complete, and separable spaces, were later called Polish spaces by N. Bourbaki.

When projective sets were introduced [C: Lusin, 1930], they became the subject of intensive investigation in Lvov. In the paper [A: Kuratowski & Tarski, 1931] the authors provide an analogy between quantifiers and actions of sets. Pursuing this line of thought further, the first author presents numerous examples of projective sets and computes their classes [A: Kuratowski, 1931].

A joint paper [A: Birnbaum & Orlicz, 1931] appears: it is the longest paper in "Studia Mathematica" prior to 1940 and contains an important theorem belonging to what would later become the theory of L^{Φ} spaces.

1932

Stanisław Mazur obtains his doctorate.

K. Kuratowski and S. Ulam travel from Lvov to attend the International Congress of Mathematicians, held in Zurich.

The monograph [A: Banach, 1932] is the first volume of a new series called Monografie Matematyczne. It makes S. Banach famous and brings international recognition to the Lvov school (see Chapter 20).

J. Schauder publishes a paper [A: Schauder, 1932b] which impresses J. Leray and marks the start of their joint work and friendship.

The work [A: Orlicz, 1932] is the start of the theory of Orlicz spaces, a theory which blossoms in the 1960s (see Chapter 21).

1933

Stanisław Ulam obtains his doctorate.

Minister Janusz Jędrzejewicz introduces major reforms to the higher education sector and abolishes some university chairs completely, including the one held by S. Ruziewicz at UJK, as well as the entire general studies department at PL, where K. Kuratowski's chair was situated. S. Ruziewicz moves to the Higher School (later the Academy) of Foreign Trade, where he joins the chair (he would teach arithmetic for business and mathematical finance), while K. Kuratowski returns to Warsaw, where he is appointed to a chair at Warsaw University.

The monograph [A: Kuratowski, 1933] appears as volume three of the Monografie Matematyczne series (following [A: Banach, 1932], [C: Saks, 1933]). For the next few decades it serves as the basic source of reference, used worldwide, on set topology. The author would continuously update it in successive editions. In 1952 he adds to it Volume II (it was written before 1939, but amazingly it was withheld).

The fourth volume of "Studia Mathematica" appears. It contains twenty-four papers, eighteen of which are from Lvov (Auerbach, two; Banach and Kuratowski, one; Banach and Mazur, two; Birnbaum and Schreier, one; Kaczmarz, two; Kuratowski, one; Mazur, one; Mazur and Orlicz, one; Mazur and Sternbach, two; Orlicz, three; Schreier and Ulam, one; Steinhaus, one).

S. Mazur and W. Orlicz generalize Banach spaces to so-called type B_0 spaces [A: Mazur & Orlicz, 1933]. In Bourbaki terminology, nowadays used quite widely, they are locally convex, complete, metrizable spaces, or, more briefly, Fréchet spaces.

Pursuing further with his interest in convexity, S. Mazur proves several theorems on linear topological spaces [A: Mazur, 1933a]. There is sometimes the accusation that the Lvov school did not fully follow through with work on general notions of linear topological spaces. Such an accusation is not entirely justifiable. S. Mazur did look at this class of spaces, because his habilitation thesis [A: Mazur, 1936] addresses, among other things, the classification of linear topological spaces and introduces convex linear topological spaces. "Because S. Mazur's results were only articulated in the abstract, in Polish, they did not have influence on the development of what is now the very important theory of linear topological spaces, and whose origins are attributed rather to J. von Neumann, J. Dieudonné, C. W. Mackey" [B: Orlicz, 1996; p. 136].

In the papers [A: Mazur, 1933a] and [A: Mazur, 1933b], the author begins intensive research into infinite-dimensional convex sets and the geometry of the unit ball in Banach spaces.

1934

After returning from Göttingen, Władysław Orlicz obtains his habilitation.

The fifth volume of "Studia Mathematica" appears. It contains twenty-one papers, fifteen of which are from Lvov (Auerbach, two; Banach and Mazur, one; Kac, one; Kaczmarz, two, Mazur and Orlicz, two; Nikliborc, one; Orlicz, three; Schauder, one; Schreier, one; Schreier and Ulam one).

1935

Herman Auerbach obtains his habilitation.

Stanisław Ulam leaves for the United States, but he would return to Lvov every year (1936, 1937, 1938, 1939) for the summer.

Mirosław Krzyżanski, from the Stefan Batory University in Vilnius, comes to Lvov for a year-long scholarship for the academic year 1935/1936.

Stefan Banach's wife, Łucja, purchases a thick exercise book for the mathematicians frequenting the Scottish Café to note down in it (rather than on the surfaces of marble tables or on scraps of paper) the problems covered in their discussions and their solutions. Stefan Banach pens in the first problem on July 17, 1935. Thus begins the famous *Scottish Book* (see Chapter 10).

Independently of J. von Neumann and A. Kolmogorov (see Chapter 20), S. Mazur introduced the notion of a locally convex space on July 9, 1935, at a talk delivered to the PTM (see his note in [A: Mazur, 1936]) and gave many results appertaining to such spaces, some of them without proofs (these were provided later by G. Birkhoff and S. Kakutani; see [B: Köthe, 1994; pp. 227–228]).

1936

Edward Otto obtains his doctorate (at PL, in descriptive geometry).

Stanisław Mazur obtains his habilitation. It was the last of six habilitations awarded in mathematics at UJK in the years 1919–1939 (Banach, 1922; Schauder, 1927; Kaczmarz, 1929; Orlicz, 1934; Auerbach, 1935; Mazur, 1936), to which one should remember the three habilitations in mathematics awarded at PL during that period (Plamitzer, 1921; Maksymowicz, 1923; Nikliborc, 1927). Such a large number of habilitations testifies to the vigor of the center and its abundance of talent.

Leaving Lvov for the International Congress of Mathematicians in Oslo are S. Banach, S. Kaczmarz, S. Mazur, J. Schauder, E. Żyliński. Banach delivers a plenary talk [A: Banach, 1937], while Mazur and Orlicz (Orlicz was absent at Oslo) and Mazur and Schauder have sessional talks. In the latter [A: Mazur & Schauder, 1936] the authors introduced the important concept of a minimalization of a functional in a reflexive space, but they did not develop it further themselves.

Jözef Marcinkiewicz arrives from Stefan Batory University (UBS) in Vilnius to spend a year-long scholarship for the academic year 1936/1937 (he obtains his habilitation degree at UBS in 1937).

The monograph [A: Kaczmarz & Steinhaus, 1936] appears as the sixth volume of the Monografie Matematyczne series. Up until the 1960s it will constitute the standard reference work on orthogonal series.

The sixth volume of "Studia Mathematica" appears. Apart from the editors (the same people as in Volume I: Stefan Banach and Hugo Steinhaus), the names of an editorial board appear on the cover: Herman Auerbach, Stanisław Mazur, Władysław Orlicz. The volume contains nineteen papers, thirteen of which are from Lvov (Auerbach, one; Eidelheit, three; Kac, one; Kac and Steinhaus, two;, Kaczmarz, one; Krzyżanski and Schauder, one; Orlicz, three; Schauder, one). The three papers of Eidelheit and the two papers of Kac herald the arrival of two great mathematical talents. However, the first of them perishes in the Holocaust just a few years later, whereas the other manages to emigrate to the United States of America. A student of H. Steinhaus, Mark Kac wrote the first of a series of papers on independent functions, the aim being a "systematic development of an idea of Steinhaus" given in [A: Steinhaus, 1923a]. Up to 1940, six papers in this series had appeared, penned either by M. Kac or Steinhaus or both of them together. After the war, four subsequent papers appear, the authors being H. Steinhaus and C. Ryll-Nardzewski.

The important paper [A: Orlicz, 1936a] appears.

1937

Mark Kac and Władysław Hetper obtain their doctorates (the latter one in logic).

The following Lvov mathematicians attend the Third Congress of Polish Mathematicians, held in Warsaw: L. Chwistek, W. Hetper, M. Kac, S. Kaczmarz, S. Ruziewicz, H. Steinhaus [B: Trzeci Zjazd, 1937], [B: Trzeci Zjazd, 1939].

Almost at the same time as A. Turing, S. Banach and S. Mazur also perceived the significance of recursive functions [A: Banach & Mazur, 1937], but unlike Turing they did not manage to develop the theme (see Chapter 22).

1938

Meier Eidelheit and Jozef Pepis obtain their doctorates.

Leon Chwistek is appointed as an ordinary professor at UJK.

Henri Lebesgue receives an *honoris causa* doctorate from UJK in May.

Mark Kac sails off in October on the "Piłsudski" to the United States of America; he will never return to Lvov again.

Hugo Steinhaus's *Mathematical Snapshots* appears, the first edition of a book that, much later on, would become very popular and will be translated into many languages.

The seventh volume of "Studia Mathematica" appears. It contains seventeen papers, nine of which come from Lvov (Auerbach, one; Banach, one; Eidelheit, one; Edidelheit and Mazur, one; Kac, two; Kac and Steinhaus, one; Kaczmarz and Marcinkiewicz, one; Schreier, one).

1939

In April Stefan Banach is made president of the Polish Mathematical Society for the period 1939–1941; in effect he will be president until his death in 1945. Stanisław Ruziewicz is appointed rector of the Academy of Foreign Trade.

In June the General Assembly of the Polish Academy of Knowledge in Cracow decides to give S. Banach an award of 80,000 zloties, the award ceremony being planned for October (the war prevented these plans and Banach never received the award).

After spending their last vacations in their homeland, Stanisław Ulam and Alfred Tarski embark on the last passage of the ship "Batory" from Gdynia to New York.

The eighth volume (the penultimate one before the outbreak of war) of "Studia Mathematica" appears. It contains eleven papers, six of which are from Lvov (Eidelheit, one; Kac, one; Kaczmarz and Turowicz, one; Nikliborc, two; Orlicz, one).

On September 1, 1939, the Germans, and then also on September 17 the Russians, invade Poland. It is the start of World War II and the nightmare years 1939–1945, a period which would see the extermination of the city of Lvov and the eradication of the Lvov School of Mathematics.

1940

Lvov is in the hands of the Soviets (see Chapter 24). UJK is closed and is replaced with the Ukrainian University. The Department of Mathematics and Physics is allocated six chairs in mathematics, five of which are occupied by Poles (S. Banach, S. Mazur, J. Schauder, H. Steinhaus, E. Żylinski), while the sixth is taken up by M. Zaryckij, who is a product of the Lvov school. Among the collaborators there are also people who fled Warsaw (B. Knaster, S. Saks, E. Szpilrajn—later Marczewski), and among the students there are J. Albrycht, A. Alexiewicz, S. Hartman, R. S. Ingarden, Z. Zahorski, and some later-known mathematicians.

One more volume of "Studia Mathematica" appears, the ninth one, and is the last one to be issued in Lvov. The editors are the same as before (Stefan Banach and Hugo Steinhaus), as is the editorial board (Herman Auerbach, Stanisław Mazur, Władysław Orlicz), but it has a newly appointed position, the technical editor, assumed by M. Zaryckij. The volume contains fourteen papers, eight of which are from Lvov (Auerbach, one; Banach, two; Eidelheit one; Erdős, Kac, van Kampen, and Wintner, one; Mazur and Orlicz, one; Steinhaus, one; Wojdysławski, one). Of the fourteen papers, eight were submitted before the war (four from abroad), four were submitted by Lvov mathematicians (Steinhaus, one; Banach, two; Wojdysławski, one) in 1940, one was sent from Równo by D. Wajnstein, and the contribution of Soviet mathematicians consisted of [C: Krein & Milman, 1940]. An indication of the new status quo is provided by the double numeration IX (I) of the volume and the inclusion of an abstract in the Ukrainian language for each paper.

Stefan Kaczmarz dies (most probably in the September campaign against the Germans), and Władysław Hepter perishes in a Gulag[108].

1941

The Higher Commission of Accreditation in Moscow confers Soviet academic titles and degrees to some Lvov mathematicians, S. Banach, S. Mazur, and J. Schauder each awarded a doctoral degree in mathematics and physics and conferred with the title of professor of mathematics.

The German-Soviet War breaks out on June 22. Lvov experiences a wave of murders. Upon entering the city, the Germans arrest twenty-three university professors, professors from the polytechnic, the Veterinarian Medical Academy, and

[108]The year of death is not known for certain. It could have been 1941 [B: Kalbarczyk, 2001; p. 100].

the Academy of Foreign Trade, together with family members staying at their homes. Almost all were then immediately shot. Of the mathematicians, the following were killed at that time: Kazimierz Bartel, Antoni Łomnicki, Włodzimierz Stożek, Stanisław Ruziewicz, Kaspar Weigel. People of Jewish extraction are in mortal danger, their only chance of survival being to stay hidden. Few can bear the strain.

Leon Chwistek escapes East and manages to reach Tbilisi.

The German occupiers shut down all the schools and colleges.

1942

In February 1942, in a building belonging to the polytechnic, the Technische Fachkurse take place, hosting activities like those at the polytechnic. A few mathematicians work there, including W. Nikliborc, K. Szałajko, M. Zaryckij.

Herman Auerbach, Ludwik Sternbach, and Menachem Wojdysławski are all killed.

1943

Meier Eidelheit and Juliusz Schauder are murdered.

1944

Marian Mojżesz Jacob is murdered. Leon Chwistek dies in Moscow.

The Soviets return and open the Ukrainian University, while at the same time the roundup of Poles commences. Stanisław Mazur is appointed leader of the Union of Polish Patriots in Western Ukraine and organizes the "repatriation" of Poles to Poland, as defined by its new borders.

1945

Stefan Banach dies and is interred in the Riedl tomb situated in the Łyczakowski Cemetery in Lvov. The last remaining Poles in Lvov are deported. Lvov becomes increasingly deserted.

The Lvov School of Mathematics ceases to exist and passes into memory.

CHAPTER 28

Chronology of Events as Perceived Elsewhere

In the book [B: Pier, 1994] there appeared a list, compiled by four authors, of the most prominant lines of research in mathematics during the years 1900–1950 (Guidelines, pp. 1–34). In it they gave, for each successive year, anything up to twenty-something names, together with their achievements (notions, theorems, theories, books, etc.). We list below the names of Polish mathematicians appearing on that list: the names of Lvov mathematicians appear in **bold font**.

1920

J. Łukasiewicz: 3-valued logic.

1922

S. Banach: complete normed vector spaces; contraction mapping theorem in complete normed spaces (see Chapter 18).
S. Banach, H Hahn: uniform boundedness theorem (see Chapter 21).

1923

H. Steinhaus: probability as a measure (see Chapter 15).

1924

S. Banach, A. Tarski: Banach-Tarski-Hausdorff paradox (see Chapter 16).
A. Tarski: finite sets.

1925

F. Leja, O. Schreier: topological groups.

1927

S. Banach, H. Steinhaus: Banach-Steinhaus theorem (see Chapter 18).
J. Schauder: extension of Brouwer fixed point theorem to Banach spaces with bases (see Chapter 22).

1929

S. Banach: "open mapping theorem" (see Chapter 21).
B. Knaster, **K. Kuratowski**, S. Mazurkiewicz: KKM lemma in fixed point theory (see Chapter 27).

1930

S. Mazur: Mazur theorem on the convex closure of compact sets.
O. Nikodym: Radon-Nikodym theorem.
R. E. A. C. Paley, A. Zygmunt: series of random functions.
H. Steinhaus: probability for the circle of convergence of an entire function to be a natural boundary (see Chapter 15).
A. Tarski: fundamental concepts of the methodology of deductive sciences.

1931

Z. W. Birnbaum, W. Orlicz: Orlicz spaces (see Chapter 21).
K. Borsuk: theory of retracts.
A. Tarski: decidability of elementary geometry.

1932

S. Banach: *Théorie des opérations linéaires* (see Chapter 20).
O. Nikodym: Dirichlet's integral as a pseudonorm.
R. E. A. C. Paley, N. Wiener, A. Zygmund: random functions.
A. Tarski: Tarski theorem on the concept of truth.

1933

K. Borsuk: antipodal theorem.
K. Kuratowski: *Topologie I*.
S. Saks: *Théorie de l'intégrale.*

1934

R. Caccioppoli, **J. Schauder**: linear elliptic equations having Hölder coefficients (see Chapter 22).
J. Leray, **J. Schauder**: degree theory in Banach spaces and method of a priori bonds (see Chapter 22).

1935

W. Hurewicz: homotopy type and homotopy groups; aspherical spaces and fundamental group; Hurewicz homomorphism.
B. Jessen, J. Marcinkiewicz, A. Zygmund: differentiation of multiple integrals on rectangular basis.
S. Kaczmarz, H. Steinhaus: *Théorie der Orthogonalreihen*[109] (see Chapter 22).
J. Schauder: functional analytic approach to hyperbolic equations (see Chapter 22).
A. Zygmund: *Trigonometrical Series.*

1936

S. Mazur, J. Schauder: minimization of functionals in reflexive spaces (see Chapter 27).

1938

S. Mazur: normed rings (see Chapter 22).

1940

J. Marcinkiewicz: Marcinkiewicz method for absolutely convergent Fourier series.

1941

W. Hurewicz: exact sequences.
W. Hurewicz, H. Wallman: *Dimension Theory.*

1942

S. Eilenberg: extensions of groups and homology.

1943

S. Eilenberg, S. MacLane: homology of groups, Bar-resolution Eilenberg-MacLane spaces.

---

[109] The monograph appeared in 1936.

1944

S. Eilenberg: singular homology.

1945

S. Eilenberg, S. MacLane: categories and functors.
S. Eilenberg, N. Steenrod: the axiomatics of homology groups of topological spaces.

1947

T. Ważewski: Ważewski topological principle for ordinary differential equations.

1948

J. G. Mikusiński: generalized distributions.
W. Szmielew: decidability of the theory of abelian groups.

1949

M. Kac: probabilistic methods in analysis and number theory.
A. Tarski: concept of essential undecidability.

Whatever one's criticisms about the above list, one can nonetheless see that world class mathematics was well represented by Polish mathematics and that Lvov had global significance in that regard. After 1940, however, Lvov vanishes from the world's mathematical map. During the war, the baton of Polish mathematics was taken up by mathematicians who had emigrated (those from Lvov included Z. W. Birnbaum, M. Kac, S. Ulam), and mathematics in Poland begins to revive in 1945, only in different centers and in a different form.

CHAPTER 29

Influence on Mathematics of the Lvov School

Let us begin with the personalities. The founders and leaders of the Lvov School of Mathematics were Hugo Steinhaus and Stefan Banach, but of their generation, a great role was also played by Antoni Łomnicki, Stanisław Ruziewicz, and Kazimierz Kuratowski, the last during the years 1927–1933. Their research and teaching talents attracted highly capable youth, who quickly made their own discoveries, thereby adding to the singular folklore status of the school. A good illustration of the rise in numbers of active members of the school can be seen through the number of doctorates and habilitation degrees, in mathematics or logic, awarded at Lvov during the years 1924–1938 (unless otherwise stated, the list below refers to degrees conferred at UJK; the list of doctorates given below is possibly incomplete):

Stefan KACZMARZ (Doctorate, 1924, thesis advisor S. Ruziewicz; habilitation, 1929).

Władysław NIKLIBORC (Doctorate, 1924, thesis advisor H. Steinhaus; habilitation, 1927, in mathematics, and habilitation, 1931, in theoretical mechanics, both habilitations obtained at PL).

Juliusz SCHAUDER (Doctorate, 1924, thesis advisor E. Żyliński; habilitation, 1927).

Sala WEINLÖS (Doctorate, 1927, thesis advisor H. Steinhaus; –).

Władysław ORLICZ (Doctorate, 1928, thesis advisor E. Żyliński; habilitation, 1934).

Zygmunt Wilhelm BIRNBAUM (Doctorate, 1929, thesis advisor H. Steinhaus; –).

Herman AUERBACH (Doctorate, 1930, thesis advisor H. Steinhaus; habilitation, 1935).

Miron ZARYCKIJ (Doctorate, 1930, thesis advisor S. Ruziewicz; –).

Stanisław MAZUR (Doctorate, 1932, thesis advisor S. Banach; habilitation, 1936).

Józef SCHREIER (Doctorate, 1932, thesis advisor S. Banach; –).

Stanisław ULAM (Doctorate, 1933, thesis advisor K. Kuratowski; –) (doctorate obtained at PL).

Marc KAC (Doctorate, 1937, thesis advisor H. Steinhaus; –).

Władysław HETPER (Doctorate, 1937, thesis advisor L. Chwistek; –) (doctorate in logic).

Meier EIDELHEIT (Doctorate, 1938, thesis advisor S. Banach; –).

Józef PEPIS (Doctorate, 1938, thesis advisor E. Żyliński; –) (doctorate in logic).

The reason why the last six people listed above did not obtain habilitation degrees is explained by the circumstances of the times: S. Ulam and M. Kac emigrated shortly after obtaining their doctorates, while the war claimed J. Schreier, W. Hetper, M. Eidelheit, and J. Pepis. For the years 1924–1938, a total of at least thirteen doctorates (possibly more) was awarded in mathematics, two in logic, together with six habilitation degrees in mathematics. These were impressive results, though Warsaw did have a slight edge on them at the time.

Thanks to the great service undertaken by S. Banach and the Lvov school in functional analysis, the notion of Banach spaces was introduced and the theory appertaining to them was developed. The significance of this contribution is best illustrated by the theorems which are fundamental to the theory and which were produced by the Lvov school, theorems such as the Hahn-Banach theorem, the Banach-Steinhaus theorem, and the theorem on inverse operators. What is more, the importance of these theorems was recognized at Lvov and was explained to a large group of eager young mathematicians, who in turn published their own results, especially in "Studia Mathematica". "One may dare to claim that at the turn of the 1930s, the world's leading center of functional analysis was in Poland [...] [and] the culminating achievement of the Lvov school was Banach's monograph" [B: Pełczyński & Semadeni, 1969; p. 86]. We further add that S. Mazur made a major contribution to the monograph by writing the closing Remarques, which "had a major influence on the development of functional analysis" [B: Köthe, 1994; p. 199]. However, numerous competitors to the Lvov school appeared in the 1930s, above all in the Soviet Union, the United States, and France. Meanwhile, in Poland, all the papers in this field were being written by at most twenty individuals, and they did not necessarily write at the same time. And there was a dramatic shortage of university positions. Furthermore, many important results obtained at Lvov were either published too late or not at all (for example, the most important aspects of type B_0 spaces, which Banach and Mazur started working on in 1933, were not published until as late as 1948 and 1953; see also A. Turowicz's comments about the Stone-Weierstrass theorem [B: Turowicz, 1995]), nonlinear operators were hardly developed (work by S. Mazur and W. Orlicz), work on the theory of Banach algebras (work by S. Mazur, M. Eidelheit, A. Turowicz) had hardly been started. The Lvov school also had its weak points [B: Pełczyński & Semadeni, 1969]:

(1) It concerned itself almost exclusively with Banach spaces defined over the real numbers, thereby neglecting spaces defined over the complex numbers, and it did not use much of the theory of analytic functions. Its links with classical analysis were also somewhat weak, which may have been due to insufficient familiarity with that subject.

(2) By concentrating on continuous operators, work on unbounded operators was neglected, meaning a large and important field of differential operators (which are noncontinuous, as a rule) was not pursued.

(3) Whereas in other parts of the world topology was advancing by virtue of the greater generality provided by the notion of an open set—i.e. a notion which does not assume the existence of a metric—work on topology in Poland tended to be confined to the metric interpretation, meaning the considered spaces had to satisfy certain separability conditions [separable spaces]. Consequently, this weakened the generality of the results obtained.

(4) It either did insufficient work or no work whatsoever on new, very important research directions in functional analysis, such as the theory of Hilbert spaces—which became a fundamental tool for quantum mechanics—or the

theory of Lie group representations, ergodic theory, almost periodic functions, etc.

Despite all these weaknesses, the Lvov School of Mathematics was nonetheless a first-rate institution of the highest class and its legacy occupies a permanent place in the history of mathematics. A nice illustration of the school's importance is the fact that the name "Banach" has, for many years, appeared more frequently in mathematics texts than any other.

However, the significance of the Lvov School of Mathematics does not rest exclusively with functional analysis. The pioneering work of A. Łomnicki and H. Steinhaus on probability theory had considerable impact. Probability was interpreted as a measure. And so, in this context, Lebesgue measure and the Lebesgue integral assumed enormous importance. As J.-P. Kahane recalls, Lvov was at that time the custodian of these notions, whereas in France they were being virtually forgotten. Also, H. Steinhaus was fortunate to have some good students, the best of them being M. Kac, with whom he wrote a long series of papers on independent functions, and thought about publishing a monograph on these functions.

Thanks to H. Steinhaus and J. Schauder, the Lvov School of Mathematics was strongly involved with classical mathematical analysis, which ensured that mathematics at Lvov did not lose its way in pure abstractions. At the same time, it attracted mathematicians with similar interests, such as A. Zygmund and J. Marcinkiewicz from Vilnius.

Another characteristic feature (shared with the Warsaw school and which may have influenced the length of K. Kuratowski's several-year period at Lvov) was the widespread use of set-theoretic and topological methods. In particular, there was widespread exploitation of the axiom of choice and Baire category methods, leading to proofs of existence theorems. Much appreciated at Lvov were elegant proofs showing the existence of "many" objects of some kind, by showing they constituted a Baire category set of type II.

In an attempt to sum things up, let us quote the opinion of a distinguished French mathematician [B: Kahane, 1995; p. 174]:

> Polish mathematics left its singular mark in every field of mathematics, distinguishing itself above all with its deployment of the axiom of choice, to establish nonconstructive existence proofs. Moreover, other, indirect constructive methods, based on Baire category theory or on probability theory (or, equivalently, on Lebesgue measure) were applied. And finally, it established functional analysis, attending to its linear aspects (Banach) and its nonlinear aspects (Schauder).
>
> In particular, Banach was partially responsible for the triumph of the axiom of choice and marked out a path to a great garden of functional analysis. Laurent Schwartz's theory of distributions became the new triumph of linear analysis. The Hahn-Banach theorem, based on the axiom of choice, appeared in all its glory in the work of the Bourbaki group. For the Bourbaki group, set theory, based on the axiom of choice, was indispensable for the very foundations upon which all mathematics rests.
>
> Such a wonderful image, as we now know, is not entirely true. Set theory achieved a new status. It is a mathematical

field like many others. It has its own axioms and these axioms can be changed. On the other hand, analysis veered to the nonlinear side, namely the domain of Schauder rather than Banach. Ultimately, thanks to computers, constructive methods now take precedence over nonconstructive ones.

Nonetheless, the legacy of Polish mathematics produced during the interwar years remains hugely important and displays a beauty that will endure through the ages.

CHAPTER 30

A Tentative Summary

The strongest field of the Lvov School of Mathematics was the theory of Banach spaces and, even more generally, functional analysis. There was the stream of outstanding papers that flowed from S. Banach and H. Steinhaus, as well as their students J. Schauder, W. Orlicz, S. Mazur, S. Ulam, M. Kac, and others. There was the special journal "Studia Mathematica", its successive volumes filled, in the main part, by their works. There was Banach's monograph and to some extent monographs of others. At the congress in Oslo, there was Banach's promotion and there was news of Lvov's interesting mathematical culture (the PTM meetings, the café life, the *Scottish Book*)—all of this made the Lvov center particularly attractive and gave it world status. Lvov was visited by the mathematical elite. Just as many could not come to Lvov in person but nonetheless took up the same set of problems and developed them further, either in new directions entirely or in the same directions pursued at Lvov.

The grandmasters of the school (H. Steinhaus and S. Banach) were at the height of their power in the 1930s. A group of young, dynamic protegés (J. Schauder, S. Mazur, W. Orlicz, and others) congregated around them, and newer members arrived. The journal "Studia Mathematica" burst into life, as did the Warsaw-Lvov series of monographs "Monografie Matematyczne"—it might have seemed that the future prospects of the school were good. Mathematics at Lvov developed strongly. However, anti-Semitism and the lack of opportunities for paid employment were already markedly evident towards the end of the 1930s. So too was the feeling that war was coming. Some talented mathematicians (Z. W. Birnbaum, M. Kac, S. Ulam) emigrated. When the threats became reality, they surpassed all expectations. World War II resulted in the total destruction of mathematics in Lvov. Most of the school's active members perished in it, and the surviving few were forced to leave the city.

Only a few active members of the Lvov mathematical school lived to see the end of the war: Stefan Banach, who survived the war feeding lice at the Weigl Institute, nonetheless died almost immediately after its conclusion; Kazimierz Kuratowski was based in Warsaw from 1934, where he helped revive Polish mathematical life; Stanislaw Mazur, after his "repatriation", stayed briefly in Łódź before settling in Warsaw, joining K. Kuratowski; Władysław Orlicz returned to Poznań after taking up an offer given to him before the war; Hugo Steinhaus, who survived the war living under an assumed name in a village near Gorlice, resettled for good in Wrocław.

Steinhaus revived "Studia Mathematica" in Wrocław in 1948. By then he was no longer involved with "operator theory", having switched his attention to applications of mathematics in the natural or technical sciences, particularly anthropology, biology, geology, medicine, agriculture, etc. He established an active school in this research area. It addressed such things as the taxonomy of Wrocław, i.e. methods for describing and classifying groups of individuals, developing important statistical tools for the natural sciences; mathematical genetics (blood types, paternity identification); the theory of system failures, including those in digger-transport-dumping systems used for mining exploration; energy forecasting for electrical power systems, etc. In this situation, "Studia Mathematica" was moved to Warsaw shortly afterwards, where it has remained ever since.

Work on functional analysis was continued by S. Mazur and W. Orlicz. The former, upon settling in Warsaw, gathered around him a group of young, talented mathematicians, including Czesław Bessaga, Aleksander Pełczyński, Stefan Rolewicz, Wiesław Żelazko, and others. W. Orlicz's first student, while he was still in Lvov, was Andrzej Alexiewicz, who followed his teacher to Poznań. Those who came to Poznań included Jerzy Albrycht, Zbigniew Ciesielski, Lech Drewnowski, Julian Musielak, Zbigniew Semadeni, Roman Taberski, as well as others. The Warsaw group continued with general problems stemming from Banach's work, while the Poznań group placed greater emphasis on specific classes of spaces, most particularly Orlicz spaces. They both were, and continue to be, important mathematical centers, recognized internationally, but their stature has never quite matched the same stature once held by the Lvov school.

CHAPTER 31

Mathematics in Lvov after 1945

After the war, as a result of Soviet repression and German atrocities, and also because of the forced deportation of Poles "across the San" [river], Lvov became ever more deserted, its population barely reaching 30,000 inhabitants. The surrounding Ukrainian countryfolk filled the vacuum. Throughout all this, the university and the polytechnic continued to function. M. Zaryckij was appointed dean of the Department of Mathematics and Physics. He himself was no longer active in research, but he did occupy himself with the task of organizing the department. The departmental chairs were filled by people from the heart of the Soviet Union. The young W. Lance was the one and only previous staff member who stayed on in Lvov. Among the newcomers were B. V. Gnedenko from Kiev (he organized a seminar on spectral theory on the basis of D. Hilbert's work and attracted some new people, but he left Lvov in 1950), A. S. Kowańko (he stayed a long time), Ju. Ja. Łopatyńskij from Baku (he did research on the theory of differential equations and algebra, and his work was of considerable influence on mathematics in Lvov after the war), L. I. Wołkovyskij (he investigated complex functions).

In January 1940, W. Lance, M. Wiszyk, and A. Strauss (1920–1999) were first year students of mathematics at the university, where they attended lectures by M. Zaryckij (analysis), Orlicz (algebra), J. Schauder (theoretical mechanics), S. Mazur (differential geometry). Moreover, they could attend seminars given by S. Banach and E. Szpilrajn (Marczewski). In the latter, W. Lance and M. Wiszyk obtained a certain result associated with problem number 192, set by B. Knaster and E. Szpilrajn (Marczewski), in the *Scottish Book* [B: Mauldin, 1981; p. 266], and it was quoted in E. Marczewski's postwar article (Fund. Math. 33 (1945)). Those three people (W. Lance, M. Wiszyk, A. Strauss) were scattered by the war. A. Strauss ended up in Uljanovsk and stayed there. M. Wiszyk returned to Lvov but moved to Moscow shortly afterwards, where he enjoyed a university career. Only W. Lance returned to Lvov and stayed there. Initially he worked (under the supervision of Jarosław Łopatyńskij) on differential equations, but he later focused on operators on Hilbert spaces and trained successors working in functional analysis. Thanks to him and his students (Professor Jaroslav Prytuła, Reader Halina Chuyko, T. Banach, O. Storoż, and others), functional analysis lives on in the mathematics at Lvov [B: Lance, Pliczko, and Storoż, 1992], [B: Chuyko, 2006].

Part VI. LIST OF LVOV MATHEMATICIANS

Mathematicians Associated with Lvov

ABAKANOWICZ-ABDANK, Bruno (1852–1900). Born in Lithuania, he studied at the polytechnic in Riga. He obtained his habilitation degree in 1876 at the technical academy in Lvov. Under the supervision of W. Żmurko, he successfully began constructing mathematical devices. He moved to Paris in 1881, where he made himself a fortune constructing various technical devices.

ALEXIEWICZ, Andrzej (1917–1995). He was born in Lvov, where he finished high school in 1937 and then took up studies at UJK, initially studying physics but transferring to mathematics after the first year. In the academic year 1944/1945 he was a laboratory assistant at the Ukrainian University. He wrote his doctorate (advisor: W. Orlicz) at the underground university in Lvov, where he defended it in 1944 (the defence of his doctoral thesis was validated at Poznań University in 1945). He was based in Poznań from 1945 onwards.

He authored fifty-six papers as well as a textbook on functional analysis.

AUERBACH, Herman (1901–1942). Born in Tarnopol, he passed his matura exam in 1919 in Lvov. He studied mathematics at UJK, and at the same time he worked there from 1923 onwards. He obtained his doctorate in 1930 and his habilitation degree in 1935. Just before his imminent deportation to the extermination camps, he killed himself by taking poison.

He authored several works on the geometry of convex figures, the theory of topological groups, etc.

BANACH, Stefan (1892–1945). Born in Cracow, he completed high school there in 1910. He studied in Lvov, first at the polytechnic there, between 1910–1914, and later at the university, between 1920–1922. He did not complete his studies. With the help of H. Steinhaus, he found employment at Lvov Polytechnic during 1920–1922 but earned his doctorate at UJK in 1920. He obtained his habilitation degree in 1922 at UJK, where he became a professor extraordinary in 1923 and an ordinary professor in 1927. From 1924 he was a correspondent member of the Polish Academy of Arts and Sciences [Polska Akademia Umiejętności]. From 1927 he was a member of the Lvov Scientific Society, while from 1931 he was a member of the Warsaw Scientific Society. In his doctoral thesis he defined a class of general spaces, the so-called *Banach spaces*, and he proved many important theorems about them, thereby initiating the theory which pertains to them. He and H. Steinhaus established the journal "Studia Matematica", publishing nine volumes between 1929–1940. The journal served as a platform for disseminating the work of the Lvov school of functional analysis and was particularly useful for the worldwide development of that theory. The monograph [A: Banach, 1932] established his fame and the importance of the Lvov school. Banach was dean of the mathematics department between 1939–1941 (he was the only Polish dean appointed at a Ukrainian university). During the German occupation his work required him to feed lice at the Weigl Institute. He was re-employed there on July 1, 1944, but died of cancer in Lvov on August 31, 1945; he is interred in the Riedle tomb in the Łyczakowski Cemetery.

He was the author of fifty-eight academic works, including three monographs, several academic textbooks, and he was the author of ten school textbooks.

BARTEL, Kazimierz (1882–1941). Born in Lvov, he completed high school there in 1901. He studied mechanics at Lvov Polytechnic between 1902–1907 and studied mathematics and philosophy at Lvov University between 1908–1909. He worked at Lvov Polytechnic from 1907, obtaining his doctorate there in 1911, as well as his habilitation degree in 1914. There he became an extraordinary professor in 1912, an ordinary professor in 1917, and served as rector for the academic year 1930/1931. He was one of the most recognizable politicians during the interwar years (he served as minister, as member of parliament, and as senator and was elected three times as the prime minister of Poland), a supporter and associate of Piłsudski. He was shot by the Germans on July 26, 1941.

He authored several works on descriptive geometry and several textbooks. His most well-known work is the monograph [A: Bartel, 1928].

BIRNBAUM, Zygmunt Wilhelm (1903–2000). Born in Lvov, he was taught at home (in Vienna during the years 1914–1918), but he completed high school in Lvov, where he passed his matura exam in 1921. He studied law at UJK, obtaining a master's degree in 1925, after which he started work in a legal practice. However, he got interested in mathematics, and between 1925–1929 he was simultaneously a student at UJK and was awarded a doctorate in 1929. He spent the following two years in Göttingen, where he gained an actuarial diploma and worked in this capacity in Vienna, but he was back in Lvov during the years 1932–1937. From 1937 onwards he was an emigré in the United States.

BÖTTCHER, Lucjan (1872–1937). Born in Warsaw, he passed his matura exam in Łomża, after which he started mathematical studies at the Russian university in Warsaw. Expelled after a year for having taken part in patriotic demonstrations, he continued his studies in Lvov and Leipzig, where he obtained a doctorate in 1898. He subsequently returned to Lvov and started employment at Lvov Polytechnic, rising through the ranks from an assistant position to readership. He retired in 1935.

He was the author of several works in mathematical analysis, a few textbooks, and a considerable number of articles on occultism, which interested him greatly.

CHWISTEK, Leon (1884–1944). He was born in Cracow, where in 1902 he completed high school and studied at UJ, where he was awarded a doctorate in 1906. He worked as a high school teacher, while during 1914–1919 he served as a legionary in the Polish army. He obtained his habilitation degree in 1928 at UJ. He was offered a position as the chair of logic at UJK in Lvov (as professor extraordinary), where he wrote and did paintings. He went to Tbilisi when the Soviet-German War broke out. He died in Moscow in 1944 (after having rejected a proposal to be minister of culture in the Polish National Liberation Committee).

He authored important works on mathematical logic. He also wrote philosophical articles and was a successful painter.

CZAJKOWSKI, Mikołaj (1887–1970). He was born in Brzeżany, where he finished high school in 1905. He studied in Prague and Vienna, where he obtained a doctorate in 1911 (advisor: F. Mertens). In 1919 he accepted a position at a chair in Kamieniec Podolski but did not stay long. A Ukrainian patriot, he taught in high schools in Lvov, Tarnopol, and Rawa Ruska. He was the director of private Ukrainian high schools in Jaworów and Rohatyn. He taught at the underground university of Lvov between 1922–1924. He spent much time and energy working for the Szewczenko Scientific Society in Lvov. He moved to Soviet Ukraine in 1929, and strived to set up a new institute of mathematics, physics, and chemistry in Odessa. Arrested in 1933, he was sent to a Gulag for ten years. Upon his release in 1943, he went to Tomsk, from where he eventually returned to Lvov as late as 1956. He was rehabilitated shortly afterwards. Now at last he could finally work at the Ukraininan University of Lvov, which he had long dreamed of. But by then he was over seventy years old and in frail health.

DZIWIŃSKI, Placyd (1851–1936). Born near Tarnopol, he finished high school in Tarnopol itself. He studied in Lvov, simultaneously attending the polytechnic and the university, finishing in 1874. He completed further studies in Berlin and Paris. He obtained his doctorate in 1881 at Lvov Polytechnic, where he also obtained his habilitation degree in 1886, then a position as professor extraordinary in 1887. From 1889 onwards he held an ordinary professorship; he was also rector for the academic year 1893/1894. He received an honorary professorship after his retirement in 1925.

He published works on geometry, algebra, and the history of mathematics, as well as very popular school and academic textbooks, including a two-volume set in 1902 and 1908.

EIDELHEIT, Meier (1910–1943). Born in Janów near Lvov, he finished high school in Lvov in 1929, as well as mathematics studies at Lvov University in 1933. He obtained his doctorate in 1938 (advisor: S. Banach). He was murdered by the Germans.

He authored several works on functional analysis.

FRANKE, Jan Nepomucen (1846–1918). He was born in Lvov, where he attended school. He studied at Lvov Polytechnic and Vienna Polytechnic. After a stay in Lvov, he studied mathematics in Zurich and Paris. From 1870 onwards, he was head of the chair of theoretical mechanics at Lvov Polytechnic, of which he was the rector during the years 1874/1875 and 1880/1881.

He published a large number of works on mechanics, including the first textbook on mechanics to be written in the Polish language, as well as on mathematics and on the history of mathematics, including a major biography of Jan Brożek.

HERCBERG, Jan (1908–1941?). Born in Warsaw, he passed his matura exam in 1927 in Częstochowa. He obtained a master's degree in mathematics at UJ in 1933. He was in Lvov from 1938, and from 1940 he was an assistant to S. Banach. After the German-Soviet War broke out he left Lvov and went missing/presumed dead.

He published a few papers on logic.

HETPER, Władysław (1909–1940?). He was born in Cracow, where he finished high school in 1927 and later obtained a master's degree in mathematics at UJ in 1932. He obtained a doctorate in 1937 at UJK in Lvov (advisor: L. Chwistek). He was captured and held by the Germans when he took part in the September Campaign. He managed to escape, but at the crossing of the San River, on the way to Lvov, he was arrested by the Soviets; he was sentenced (apparently he was carrying with him papers on logic, and the Soviets mistook the mysterious symbols therein for a spy's secret code) and sent to prison. He died in a Gulag.

He wrote several papers on logical semantics. He was regarded as the person who would have given Chwistek's ideas a more mathematical treatment.

JACOB, Marian Mojżesz (1900–1944). Born in Nadwórna, he spent his youth in Palestine. He studied mathematics in Vienna, where he obtained a doctorate in 1925. He worked as an actuary in Trieste and Warsaw. He spent the years 1939–1944 in Lvov, first at the Ukrainian University, then at the Todt Institute. After leaving for Warsaw, he perished under unknown circumstances.

He published papers on insurance mathematics and on orthogonality.

JANISZEWSKI, Zygmunt (1888–1920). Born in Warsaw, he attended high school in Lvov. During 1907–1911 he studied mathematics in Zurich, Göttingen, Paris, and Munich. He obtained a doctorate in 1911 at the Sorbonne, and his habilitation degree in Lvov, after Sierpiński had asked him to join him there. He spent the first winter of war as a volunteer legionary (sergeant) on the Carpathian front. He was made a professor extraordinary at Warsaw University in 1919. He died in 1920 in Lvov.

The author of several papers on topology, he devised a development program for Polish mathematics and was a cofounder of "Fundamenta Mathematicae".

KAC, Mark (1914–1984). He was born in Krzemieniec, where he completed the famous Lycée in 1931. He studied mathematics in Lvov at UJK, obtaining a doctorate in 1937 (thesis advisor: H. Steinhaus). A year later he emigrated to the United States, where he enjoyed an immensely successful academic career.

He was the author of nearly 150 papers and books, principally on stochastic processes and statistics.

KACZMARZ, Stefan (1895–1940). Born in Sambor, he studied mathematics at UJ between the years 1915–1922, with an interlude for doing his military service as a legionary in the war. He was in Lvov from 1923, where he obtained a doctorate in 1924 at UJK and a habilitation degree in 1929. He was an assistant professor (adjunct) and a reader at Lvov Polytechnic. As a lieutenant in the reserves, he took part in the September Campaign, from which he never returned. The circumstances of his death are uncertain.

He was the author of thirty-two works in functional analysis and the theory of orthogonal series, including the monograph [A: Kaczmarz & Steinhaus, 1936].

KĘPIŃSKI, Stanisław (1867–1908). Born in Bochnia, he passed his matura exam in 1885 in Cracow, where he studied mathematics, physics, and astronomy at UJ between 1885–1889. Upon obtaining his doctorate in 1890, he went on to postdoctoral studies in Göttingen, from where he returned to do his habilitation degree at UJ in 1893. He lectured mathematics at UJ (from 1896 as professor extraordinary), and in 1898 he took up a chair in mathematics in Lvov at the polytechnic school, of which he was rector for the academic year 1903/1904. He also lectured at Lvov University.

He was the author of the highly regarded *Podręcznik równań różniczkowych* [Textbook on Differential Equations] and of several works on the theory of differential equations.

KNASTER, Bronisław (1893–1980). He was born in Warsaw, where he finished high school in 1911. He started studies in medicine at the Sorbonne in Paris but never completed them because World War I forced him to stay in his homeland. He took up the study of mathematics at the newly revived Warsaw University, which—after an interlude while taking part in the Russo-Polish War—rewarded him with a doctorate in 1923 (advisor: S. Mazurkiewicz) and a habilitation degree in 1926. He worked at Warsaw University as an adjunct and reader. He spent the years 1939–1945 in Lvov, first at the Ukrainian University (where he was nominated for a professorship), then at the Weigl Institute, where he worked as a lice feeder. He was based in Wrocław after the war.

He was the author of some fifty papers in topology and the theory of sets.

KRYGOWSKI, Zdzisław (1872–1955). Born in Lvov, he finished his schooling in 1890 in Cracow, where he studied mathematics at UJ between 1890–1895, ending up with a doctorate. He completed his studies in Berlin and Paris during the years 1895–1898. He lectured mathematics at the polytechnic in Lvov from 1901, obtained his habilitation degree in 1907, and during the period 1908–1919 he was a professor there and director of a mathematics chair. He was the rector there for the academic year 1907/1908. He accepted an offer in 1919 to take up a mathematical chair at the newly established Poznań University, and from that time onwards he was linked with Poznań.

He was the author of nearly twenty works, most in the Polish language, including two books.

KURATOWSKI, Kazimierz (1896–1980). Born in Warsaw, where he finished high school in 1913, he started mathematics studies at Glasgow University. From 1915 he continued his studies at Warsaw University. There he obtained his doctorate in 1921 and his habilitation degree immediately afterwards. During 1927–1933 he was a professor at the PL general studies department and subsequently a professor at Warsaw University. He was an active organizer of mathematical life.

He was the author of the highly regarded monograph *Topologie* [C: Kuratowski, 1952; first ed., 1933]; he then focused on extending it. In total, he wrote nineteen monographs and academic textbooks in analysis, the theory of sets, and topology; 172 papers on analysis, the theory of sets, and topology; as well as thirty-five articles of personal reminiscences and of a historical nature. Selected works of his appeared as [A: Kuratowski, 1988].

LANCE, Władysław (1920–2007). Born in Warsaw, he attended school in Lvov. In January 1940 he started studies in mathematics at the Ukrainian University there. He struggled through the war years 1941–1944 in Bashkiria, finishing his studies in Lvov between 1945–1948. After postgraduate work he obtained his candidate of science degree in 1957 (equivalent to a Polish doctorate), and in 1964 he obtained his doctoral degree (equivalent to a Polish habilitation degree). He remained in Lvov, first employed at the polytechnic, and as an Academy of Science postgraduate, he moved to the university in 1965, where he became a professor in 1967 and director of a mathematical chair.

He was the the author of over 130 works on the theory of differential equations, the theory of spectral operators, and nonstandard analysis, as well as translations, including translation from Polish into Russian of K. Maurin's *Hilbert Space Methods* (Moscow, 1965).

LEJA, Stanisław (1911–2000). He held a master's degree in mathematics and was deputy assistant at UJK. Imprisoned in the heart of the Soviet Union, after the "amnesty" of 1941 he ended up in the United States, where he became a professor at the University of Kalamazoo in New Jersey.

LEMOCH, Ignacy Wojciech (1802–1875). He was born in Netvorice (now in the Czech Republic). Nothing is known about his school days. In 1840, when he was a building consultant in Vienna, he was appointed to a professorship in applied geometry at the university in Lvov, where he remained active until 1870 teaching elementary and higher mathematics. He was the dean of the philosophy deparment three times and was rector for the 1854/1855 term. He was one of the organizers of the Technical Academy in Lvov and later a lecturer there.

He was the author of the textbook *Lehrbuch der praktischen Geometrie* (1849).

LEWICKIJ, Włodzimierz (1872–1956). Born in Tarnopol, he attended schools in Złoczowa, Tarnopol, and Lvov. During 1890–1894 he studied at the university in Lvov. A student of J. Puzyna, he and S. Dickstein corresponded with each other by post. In 1894 he published his first academic mathematical work in the Ukrainian language. After receiving his doctorate in 1901, he left for further study in Germany. He taught at the Fifth High School in Lvov and wrote school textbooks in Ukrainian. A Ukrainian patriot, he lectured at the underground Ukrainian University after the official one had been closed down. He was an outstanding activist for the Ukrainian Science Society, named after T. Szewczenko. He became the director of the chair of mathematical analysis at the university in Lvov in 1945, and he tried to preserve the tradition of the Lvov school of functional analysis.

He was the author of thirty works in the theory of analytic functions.

ŁOMNICKI, Antoni Marian (1881–1941). Born in Lvov, he studied mathematics at the university in Lvov during 1899–1903, then completed them in Göttingen between 1906–1907. He was a high school teacher during 1907–1919. He obtained his habilitation degree in 1919 at Lvov Polytechnic, where he worked between 1920–1941, first as a professor extraordinary and then from 1921 onwards as an ordinary professor and director of a mathematical chair. He was shot by the Germans on July 4, 1941.

He was the author of over thirty works, mainly on probability theory and mathematical cartography, in addition to eighteen books, including fifteen school textbooks. Even today, his monographs [A: Łomnicki, 1927], [A: Łomnicki, 1956] are regarded as among the finest ever presentations of mathematical cartography.

ŁOMNICKI, Zbigniew (1904–1994). He was the nephew of Antoni. Born in Kołomyja, he attended schools in Vienna, Morawia, Cracow, and Lvov. After his matura exam, he studied mathematics and physics at UJK during 1922–1926. He was involved with insurance and worked in Warsaw from 1931. He spent the war years in the army, earning the highest British and Polish military commendations, and attained the rank of captain. He remained abroad as an emigré. He published over thirty papers on statistics, insurance mathematics, and stochastic processes, including joint papers with S. Ulam and S. K. Zaremba.

ŁUKASIEWICZ, Jan (1878–1956). He was born in Lvov, where he finished a philology school in 1897 and then started studies in Lvov at the university. He initially chose law but quickly switched to philology, while also attending mathematics lectures. His thesis advisor (doctorate in 1902) and teacher was Kazimierz Twardowski, founder of the Lvov School of Philosophy. He spend the year 1905/1906 completing his studies in Berlin and Louvain. Upon his return, he obtained his habilitation degree in 1906 at the university in Lvov, where he became a professor in 1911. In 1915 he accepted an offer at the newly revived Warsaw University, assuming a chair in philosophy. Since he knew a lot of mathematicians, his influence on them was considerable. He left Warsaw in 1944 and lived abroad thereafter.

MAZUR, Stanisław (1905–1981). He was born in Lvov, where he passed his matura exam and studied mathematics at UJK. He never completed his studies. During 1926–1932 he worked at UJK, obtaining a doctorate in 1932. He worked at Lvov Polytechnic between 1935–1939, obtaining his habilitation degree in 1936. He was a professor and a head of a chair of mathematics at the Ukrainian University. He left Lvov in 1946, reaching Łódź (he declined an offer to go to Wrocław), from where he moved to Warsaw in 1948. He was a cofounder of the Lvov school of functional analysis and its continuation in Warsaw.

He produced nearly forty papers in functional analysis, many of fundamental importance. He had a reluctance to publish and some of his results were later obtained a second time.

MAZURKIEWICZ, Stefan (1888–1946). Born in Warsaw, where he finished high school in 1906, he passed his matura exam in Cracow. He studied in Cracow, Lvov, Munich, and Göttingen. He obtained his doctorate in 1913 in Lvov and his habilitation degree in 1919 at UJ. He was a professor extraordinary at Warsaw University from 1919 and ordinary professor from 1920.

He was the author of 141 works, including several books. He was a member of the Polish Academy of Arts and Science, and he was president of the Polish Mathematical Society in the years 1932–1936 and 1937–1939.

MOROŃ, Zbigniew (1904–1971). Born in Jarosław, he finished high school in Drohobycz in 1923. He studied at UJK between 1923–1929; afterwards he taught in schools in various places around Poland. He spent the war in Drohobycz, and after it he settled in Wrocław.

He was known as the author of a student work [A: Moroń, 1925].

NIKLIBORC, Władysław (1899–1948). He was born in Wadowice, where he passed his matura exam in 1916. He studied mathematics at UJ between 1916–1922, with an interlude period in which he fought (legionary, Austrian army, Russo-Polish War). He spent the period 1922–1932 working at Lvov Polytechnic. He obtained his doctorate in 1924 at UJK, where he obtained his habilitation degree in mathematics in 1927. He did postdoctoral study in Dresden, Leipzig, Göttingen, and Paris. After his habilitation degree in theoretical mechanics, obtained in 1931 at Lvov Polytechnic, he became a reader there in mathematics and theoretical mechanics. He was a professor extraordinary at Warsaw Polytechnic during 1937–1939, but he spent the war years in Lvov as a professor at the Ukrainian University and German Technical Courses. He was back in Warsaw from 1945. He committed suicide following an overnight interrogation by the secret police.

He was the author of twenty-seven papers in classical analysis, differential equations, the calculus of variations, theoretical mechanics, hydromechanics, etc.

ORLICZ, Władysław (1903–1990). Born in Okocim, he attended school in Tarnopol and Lvov, where he first studied mechanics at the polytechnic and then mathematics at UJK. He completed his studies in 1926 but was an assistant at UJK from 1924. There he obtained his doctorate in 1928 (advisor: Zyliński) as well as his habilitation degree in 1934. He worked at Lvov Polytechnic during 1931–1937, obtaining a professorship at Poznań University in 1937. He spent World War II in Lvov but was back in Poznań after 1945. He set up a school in Poznań and had many students.

He produced 171 papers in functional analysis, a monograph [C: Orlicz, 1992], and two books.

OTTO, Edward (1908–1986). Born in Lvov, he passed his matura exam there in 1927 and studied at the general studies department of Lvov Polytechnic between 1927–1932, specializing in descriptive geometry and obtaining a master's degree. He obtained his doctorate in 1936 (advisor: A. Plamitzer), then moved to Warsaw Polytechnic in 1937, where he stayed for good. There he obtained his habilitation degree in 1945 and became a professor extraordinary in 1946 and an ordinary professor in 1961. He worked for the Polish Mathematical Society and was involved with mathematical olympiads.

He authored several dozen works on descriptive and projective geometry.

PEPIS, Józef (1910–1943?). Born in Tarnopol, finishing his schooling there in 1928, he studied mathematics at UJK between 1928–1933, ending up with a master's degree. He had difficulty finding employment, even in schools. He obtained a doctorate at UJK in 1938. He was a reader at the Ukrainian University from 1939. He perished in the war under unknown circumstances.

He published a few worthy works in mathematical logic, concerning decidability of a narrower predicate calculus (author of the "Pepis prefix").

PLAMITZER, Antoni Karol (1889–1954). He was born near Cracow. He studied at the polytechnic in Lvov between 1908–1914, and he studied mathematics at the university in Lvov between 1910–1913. He obtained his doctorate in 1914 at the polytechnic in Lvov, where he also obtained his habilitation degree in 1921—in both cases specializing in descriptive geometry. He was made a professor extraordinary in 1922 at Lvov Polytechnic (ordinary professorship from 1927). He was in Silesia from 1945.

He was the author of sixteen papers on descriptive and projective geometry and the author of three textbooks.

PUZYNA, Józef, Prince of Kozielsk (1856–1919). Born in Nowy Martynów (Eastern Galicia), he finished his schooling in Lvov in 1875, then studied mathematics at the university in Lvov. After obtaining his doctorate in 1883, he did postdoctoral work at Berlin University between 1883–1884 under K. Weierstrass among others. He then worked for the rest of his life at the university in Lvov, after obtaining his habilitation degree there in 1885, where he became an ordinary professor in 1892 and head of a chair of mathematics. He served as dean and rector. He was a member of the Academy of Arts and Sciences.

He was the author of twenty-two works, several of them in German, as well as what was an excellent monograph for its time [A: Puzyna, 1898–1900]. He was a superb and versatile lecturer and a great lover of music.

RAJEWSKI, Jan (1857–1906). Born in Małopolska, he finished his schooling in 1875 in Lvov, where he also studied and obtained a doctorate in 1884. He was a professor extraordinary at the university in Lvov from 1900.

He was the author of several papers on the theory of differential equations.

RUZIEWICZ, Stanisław (1889–1941). He was born in Podstaje, near Kołomyja. He finished his schooling in Kołomyja in 1908. He studied at the university in Lvov between 1908–1913, where he obtained his doctorate in 1912 (advisor: J. Puzyna). He subsequently spent a year in Göttingen. He obtained his habilitation degree at the university in Lvov in 1918 and was appointed as a professor extraordinary there in 1920, becoming an ordinary professor in 1924. He lost his chair at UJK in 1934 as a result of educational reforms introduced by Jędrzejewicz. He transferred to the Academy of Foreign Trade, of which he became rector in 1939. He was shot by the Germans on July 12, 1941.

He was the author of thirty-four papers on the theory of functions and the author of two textbooks.

SCHAUDER, Juliusz Paweł (1899–1943). He was born in Lvov, where he finished his schooling in 1917, just before he was taken into the Austrian army and sent to the Italian front. Having been taken prisoner there, he volunteered to join the Polish army in France. Returning home in 1919, he studied mathematics at UJK between 1920–1923, where he gained his doctorate in 1924 and his habilitation degree in 1927. He lectured at UJK, but school teaching was his principal means of earning a living. He was made a professor at the Ukrainian University in 1939 in Lvov and worked at the Ukrainian Academy of Science. When the Germans came, he hid in Borysław but returned to Lvov in 1943. He found it extremely difficult to live a life of hiding. He was shot in the street by a German patrol.

He was the author of thirty-three papers on integration theory, the topology of Banach spaces, and second-order partial differential equations, the initiator of nonlinear functional analysis.

SCHREIER, Józef (1909–1943). Born in Drohobycz, he studied mathematics at UJK in Lvov. He obtained his doctorate at UJK. He worked as a school teacher and was a friend of Ulam. He published several works on topological groups, permutation groups, set theory, and algebra. He was a chess devotee. He had to go into hiding when the Germans arrived. He took poison after being betrayed.

SIERPIŃSKI, Wacław (1882–1969). He was born in Warsaw, where he finished his schooling in 1900 and where he studied at the Imperial University until the strike of 1904. He continued his studies in Cracow, where he obtained his doctorate in 1906, and in Göttingen. He obtained his habilitation degree at the university in Lvov in 1908, becoming a professor extraordinary there in 1910. He established an active mathematical group there, but it broke up due to his internment in Russia between 1914–1917 and because of his acceptance of a professorship at Warsaw

University. He was a frequent visitor to Lvov during the period 1918–1939, which bore fruit in the form of numerous joint papers with Lvov mathematicians and jointly written school textbooks. He was a member of the Academy of Arts and Sciences and the Polish Academy of Sciences.

He was the author of over 720 works, including many books. He had a major influence on mathematics in Poland and considerable influence on mathematics elsewhere around the world.

STARK, Marceli (1908–1971). He was born in Lvov, where he finished school. He studied mathematics at UJK, obtaining a master's degree in 1933. He worked only briefly as a mathematics assistant at UJK because he was suspected of illegal communist activities. After the Soviets came, he again became a mathematics assistant, but he had to hide once the Germans arrived. Arrested in 1943, he was sent to several concentration camps. He survived. He was an adjunct in Wrocław and moved to Warsaw in 1951, where he involved himself with editorial roles.

He was the author or coauthor of several academic textbooks.

STEINHAUS, Hugo (1887–1972). Born in Jasło, where he finished his schooling in 1905, he started mathematics studies at the university in Lvov. He moved to Göttingen after his first year, where he obtained his doctorate summa cum laude in 1911. He returned to his homeland, served in the Polish legions in the Wołyń Campaign of 1915, and then worked in both Cracow and Lvov for the Central Office of Country Reconstruction. He obtained his habilitation degree in 1917 at the university in Lvov, where he became a professor extraordinary in 1920 and an ordinary professor in 1923. The period he spent in Lvov between 1920–1941 was the most creative time of his professional life. He helped Banach get settled into Lvov, and not long afterwards a capable group of young mathematicians started clustering around them. He and Banach founded "Studia Mathematica", and between 1929–1940 they published nine volumes of the journal, which became a disseminating platform for the work of the Lvov school of functional analysis. He spent the period 1941–1945 in hiding, first in Lvov and then in a village near Gorlice, where he gave lessons on a clandestine basis. He accepted a chair in mathematics in Wrocław after the war in 1945. He remained at Wrocław University for the rest of his life.

He was the author of over 250 papers in many areas of mathematics and several books popularizing mathematics, the most famous one being *Mathematical Snapshots* [A: Steinhaus, 1938b] (four editions in Polish, translated into several dozen languages), undoubtedly one of the best internationally known textbooks on mathematics written by a Polish author.

STERNBACH, Ludwik (1905–1942). He was born in Sambor, where he finished his schooling in 1923. He studied mathematics and physics at UJK between 1923–1928 and later worked as a school teacher. He published two papers in 1933 (jointly with Mazur). He worked for an insurance society in Warsaw between 1934–1939. Upon returning to Lvov, he became an assistant at the Ukrainian University, but he had to go into hiding when the Germans arrived. The circumstances of his death remain unknown.

STOŻEK, Włodzimierz (1883–1941). Born near Cracow, he studied at UJ between 1876–1880. He spent the period 1906–1908 in Göttingen. He obtained his doctorate at UJ in 1922. In the same year he became a professor extraordinary at

Lvov Polytechnic and became an ordinary professor there in 1926. He was dean several times. He was shot by the Germans on July 4, 1941.

He was the author of several works on analysis and the coauthor of several school textbooks.

TUROWICZ, Andrzej (1904–1989). Born in Przeworsk, he attended school in Cracow, where he passed his matura exam in 1922. He studied mathematics at UJ between 1922–1928 and taught mathematics in schools until 1937. He spent 1937–1939 as an adjunct at Lvov Polytechnic, becoming a lecturer there when the Soviets arrived. He returned to Cracow in 1941 and took Holy Orders as a Benedictine monk in Tyniec in 1945. He obtained his doctorate in 1946 and worked at UJ. He transferred to the Mathematical Institute of the Polish Academy of Sciences in 1962, obtaining his habilitation degree there in 1963. He became a professor extraordinary in 1969.

During his Lvov days he worked together with S. Kaczmarz (they wrote a joint work on the form of a multiplicative functional) and S. Mazur. After the war he produced several dozen papers on differential equations, functional analysis, the theory of stereography, probability theory, etc.

ULAM, Stanisław Marcin (1909–1984). He was born in Lvov, where he finished his schooling in 1927. He studied mathematics at the general studies department, Lvov Polytechnic, where he obtained his doctorate in 1933. He worked at Vienna University, Zurich Polytechnic, and Cambridge University. He was at Princeton University from 1935 but spent the summers in his homeland. While he was in Lvov, he worked closely with S. Mazur and J. Schauder, later also with Kac. He was interested in set theory, topology, functional analysis, probability theory, etc. During the war he played a major role in the Manhattan Project. He visited Poland a couple of times after the war.

He published over 100 papers and several books, mostly in the United States.

WOJDYSŁAWSKI, Menachem (1918–1942). He was born in Łódz, where he finished the Second Jewish School in 1936, subsequently studying mathematics at Warsaw University. He obtained a master's degree in 1939 at the Ukrainian University in Lvov and worked there as an assistant and later as a postgraduate. The circumstances of his death are unknown.

He was the author of three papers on topology.

ZAJĄCZKOWSKI, Władysław (1837–1898). He was born in Strzyżów, near Rzeszów. He attended schools in Cracow and Rzeszów. He studied mathematics and physics at UJ between 1855–1859, obtaining a doctorate there in 1861. After postdoctoral work in Göttingen, Berlin, and Vienna, he obtained his habilitation degree in 1862 at UJ. Between 1865–1872 he lectured mathematics in Warsaw, first at the Main School, later at the Imperial University. He was an ordinary professor at Lvov Polytechnic from 1872 and was rector there in 1878/1879 and 1885/1886. He was a member of the Academy of Arts and Science in Cracow and a correspondent member of the Academy of Sciences in Paris.

He published dozens of papers in mathematical analysis and several textbooks, as well as a history of Lvov Polytechnic (up until 1894).

ZARYCKIJ (Zarzycki) Miron (1889–1961). He was born in Mogilnica, near Trembowla. He attended schools in Brzeżany, Tarnopol, and Przemyśl. After his matura exam in 1907, he started studying at Vienna University but transferred to Lvov after the first year, where he was taken under Sierpiński's wing. Completing his studies in 1912, he worked as a school teacher in Lvov, Bełz, Kołomyja, and Tarnopol. At the urging of Steinhaus, he moved to Lvov in 1925, where he worked in a school and began research work, obtaining a doctorate at UJK in 1930. He was very active in the T. Szewczenko Scientific Society in Lvov. When the Soviets took over Lvov, he was made the vice dean of the Department of Mathematics and Physics at the Ukrainian University in Lvov and was made the dean between 1945–1947; at the same time he lectured at the polytechnic in Lvov, and between 1950–1955 he organized a mathematics chair at the university in Użgorod. He and his daughter Katarina (1914–1986) have a street named after them in Lviv (before the war it was called Obertyńska Street).

He published nearly twenty papers on the theory of sets and the algebra of logic, together with nearly twenty popular science articles on mathematics, mostly of a historical nature.

ŻMURKO, Wawrzyniec (1824–1889). Born in Przemyśl, he finished high school in 1844 in Przemyśl. He studied law at the university in Lvov during the academic year 1845/1846 and then scientific subjects at the polytechnic and university in Vienna, simultaneously. He was a reader at Vienna Polytechnic from 1849. He moved to the polytechnic in Lvov in 1851 and was an ordinary professor at the university in Lvov and a head of a chair at the university in Lvov.

He was the author of twenty-six works, including *Wykłady Matematyki* [Mathematical Lectures] (1864) and other textbooks. He was known as a constructor of mathematical devices.

ŻORAWSKI, Kazimierz (1866–1953). He was born in Szczurzyn, near Ciechanów. He completed his schooling and higher studies in Warsaw, obtaining in 1888 the degree of candidate of mathematics at the Imperial University. He managed to get to Germany to do further studies, where S. Lie had the greatest influence on him and under whose supervision he obtained his doctorate (Leipzig, 1891). He obtained his habilitation degree in 1892 at UJ, but he started work at the Polytechnic School in Lvov, where he was an assistant and private reader until 1895 (there was no reply from the school regarding his application for a position as professor extraordinary). He became a professor extraordinary at UJ in 1895, was an ordinary professor there from 1898 onwards, and was the rector during the academic year 1917/1918 (the original intention was to appoint M. Smoluchowski, but that never happened because the latter died prematurely). He was an ordinary professor in Warsaw from 1919 at Warsaw Polytechnic, and from 1926 he was an ordinary professor at Warsaw University.

He was the author of nearly fifty papers and a pioneer of differential geometry in Poland.

ŻYLIŃSKI, Eustachy (1889–1954). Born near Winnica (Ukraine), he finished high school in Kiev in 1907 and completed university studies at Kiev University in 1911. Receiving a grant to become a university professor, he went away in 1912/1913 to complete his studies in Göttingen, Marburg, and Cambridge, after which he obtained a master's degree in 1914. He served in the Russian army until 1917, first completing officer school in Kiev and Petersburg and then lecturing on electrotechnical subjects to military staff stationed at the southwest front at

Kamieniec Podolski and Berdyczów. During the period 1917–1919 he was a reader at the Polish university college and the Ukrainian State University in Kiev and then again lectured in the Polish army on electrotechnics in Warsaw. In autumn 1919 he was appointed professor extraordinary at the university in Lvov, where he became an ordinary professor in 1922. He participated in mathematical congresses at Cambridge, Bolonia, and Oslo. During the years 1939–1941 and 1944–1946 he was a professor at the Ukrainian University in Lvov. He settled in Łódź after the war and worked in Gliwice.

He authored several papers and a couple of textbooks.

Also studying at the Polish university in Lvov were others who later became well-known mathematicians and logicians, such as Jerzy ALBRYCHT, Ludwik BORKOWSKI (1914–1991), Franciszek LEJA (1885–1979), Otton NIKODYM(1887–1974), Eustachy TARNAWSKI (1903–1992), but their academic work progressed outside Lvov.

Part VII. BIBLIOGRAPHIES

A. Mathematical Works by Lvov Mathematicians[110]

Herman AUERBACH and Stefan BANACH

[1931] *Über die Höldersche Bedingung*, Studia Math. 3, pp. 180–184; reprint: [A: Banach, 1967; pp. 223–227].

Stefan BANACH

[1920] *Sur l'équation fonctionnelle* $f(x + y) = f(x) + f(y)$, Fund. Math. 1, pp. 123–124; reprint: [A: Banach, 1967; pp. 47–50].

[1922a] *Sur les opérations dans les ensembles abstraits et leurs applications aux équations intégrales*, Fund. Math. 3, pp. 133–181; reprint: [A: Banach, 1979; pp. 305–348].

[1922b] *An example of an orthogonal development whose sum is everywhere different from the developed function*, Proc. London Math. Soc. 2(21), pp. 95–97; reprint: [A: Banach, 1967; pp. 63–65].

[1923] *Sur le problème de mesure*, Fund. Math. 4, pp. 7–33; reprint: [A: Banach, 1967; pp. 66–89].

[1929a] *Sur les fonctionnelles linéaires*, Studia Math. 1, pp. 211–216 and 223–239; reprint: [A: Banach, 1979; pp. 375–395].

[1929b] *Teorja operacyj*, tom I: *Operacje linjowe*, Warszawa: Kasa im. Mianowskiego.

[1930a] *Über additive Massfunktionen in abstrakten Mengen*, Fund. Math. 15, pp. 97–101; reprint: [A: Banach, 1967; pp. 200–203].

[1930b] *Über einige Eigenschaften der lakunären trigonometrischen Reihen*, Studia Math. 2, pp. 207–220 and 251; reprint: [A: Banach, 1967; pp. 187–199].

[1931a] *Über metrische Gruppen*, Studia Math. 3, pp. 101–113; reprint: [A: Banach, 1979; pp. 402–411].

[1931b] *Die Baire'sche Kategorie gewisser Funktionenmengen*, Studia Math. 3, pp. 174–179; reprint: [A: Banach, 1967; pp. 218–222].

[1931c] *Teorya operacyj liniowych*, [1932] *Théorie des opérations linéaires*, Monografie Mat. 1, Nowum, Warszawa.

[1932] *Théorie des opérations linéaires*, Monografie Matematyczne 1, Warszawa; reprint: [A: Banach, 1967; pp. 13–219]; translations: Ukrainian (Kiev, 1948, translated by M. Zaryckij), English (North-Holland, 1987; Chelsea, n.d.).

[110] Works by mathematicians who were associated with Lvov on a temporary or occasional basis (W. Sierpiński, K. Kuratowski, and others) are listed depending on whether the authors in question either wrote them in Lvov or were officially based there at the time of writing.

[1937] *Die Theorie der Operationen und ihre Bedeutung für die Analysis*, Comptes Rendus du Congrès International des Mathématiciens (Oslo, 1936): A. W. Brøggers Bokktrykkeri, pp. 261–268; reprint: [A: Banach, 1979; pp. 434–441].

[1938a] *Über homogene Polynome in* (L^2), Studia Math. 7, pp. 36–44; reprint: [A: Banach, 1979; pp. 442–449].

[1938b] *Mechanika*, 2 vols., Monografie Matematyczne 8 and 9, Warszawa; revised Polish edition: 2 vols., Warszawa: Czytelnik, 1947; 4 ed. with changes, Biblioteka Matematyczna 13, Warszawa: Państwowe Wydawnictwo Naukowe, 1956; English translation: *Mechanics*, translated by E. J. Scott, Monografie Matematyczne 34, Warszawa-Wrocław 1951.

[1951] *Wstęp do teorii funkcji rzeczywistych* (Introduction to real-valued functions), Monografie Matematyczne 17, Warszawa.

[1967] *Oeuvres I*, Warszawa: Państwowe Wydawnictwo Naukowe.

[1979] *Oeuvres II*, Warszawa: Państwowe Wydawnictwo Naukowe.

Stefan BANACH and Kazimierz KURATOWSKI

[1929] *Sur une généralisation du problème de la mesure*, Fund. Math. 14, pp. 127–131; reprints: [A: Banach, 1967; pp. 182–186] and [A: Kuratowski, 1988; pp. 327–331].

Stefan BANACH and Stanisław MAZUR

[1937] *Sur les fonctions calculables*, Ann. de la Soc. Polon. de Math. 16, p. 223.

Stefan BANACH and Stanisław RUZIEWICZ

[1922] *Sur les solutions d'une équation fonctionnelle de J. Cl. Maxwell*, Bulletin International de l'Académie Polonaise des Sciences de Cracovie, Classe des Sciences Mathématiques et Naturelles, Série A, pp. 1–8; reprint: [A: Banach, 1967; pp. 51–57].

Stefan BANACH and Hugo STEINHAUS

[1918] *Sur la convergence en moyenne de séries de Fourier*, Bulletin International de l'Académie des Sciences de Cracovie, Année 1918, Classe des Sciences Mathématiques et Naturelles, Série A: Sciences Mathématiques, pp. 87–96; reprints: [A: Banach, 1967; pp. 31–39] and [A: Steinhaus, 1985; pp. 215–222]

[1927] *Sur le principe de la condensation de singularités*, Fund. Math. 9, pp. 50–61; reprints: [A: Banach, 1979; pp. 365–374] and [A: Steinhaus, 1985; pp. 363–372].

Stefan BANACH and Alfred TARSKI

[1924] *Sur la décomposition des ensembles de points en parties respectivement congruentes*, Fund. Math. 6, pp. 244–277; reprints: [A: Banach, 1967; pp. 118–148] and [C: Tarski, 1986; vol. I, pp. 119–154].

Kazimierz BARTEL

[1928] *Perspektywa malarska* (Perspective in Painting), tom I, Lwów: Książnica-Atlas; revised: tom I, Warszawa: Państwowe Wydawnictwo Naukowe, 1955;

tom II, Warszawa: Państwowe Wydawnictwo Naukowe, 1958; translation: Kazimierz Bartel, *Malerische Perspektive, Grundsätze, geschichtlicher Überblick, Ästhetik*, hrsg. von Wolfgang Haack, Band I, Leipzig-Berlin, B.G. Teubner, 1933.

Zygmunt BIRNBAUM and Władysław ORLICZ

[1931] *Über die Verallgemeinerung des Begriffes der zueinander konjugierten Funktionen*, Studia Math. 3, pp. 1–67; reprint: [A: Orlicz, 1988; vol. I, pp. 133–199].

Meier EIDELHEIT

[1936a] *Zur Theorie der konvexen Mengen in linearen normierten Räumen*, Studia Math. 6, pp. 104–111.

[1936b] *Zur Theorie der Systeme linearer Gleichungen*, Studia Math. 6, pp. 139–146.

[1938] *Zur Theorie der Systeme linearer Gleichungen (II)*, Studia Math. 7, pp. 150–154.

[1940] *On isomorphisms rings of linear operators*, Studia Math. 9, pp. 97–108.

Zygmunt JANISZEWSKI

[1911] *Sur les continus irréductibles entre deux points*, Thèse, Paris: Gauthier-Villars; reprints: Journal de l'École Polytechnique 16 (1912), pp. 79–170, and [A: Janiszewski, 1962; pp. 31–125].

[1912] *Über die Begriffe "Linie" und "Fläche"*, Intern. Congress of Mathematicians, Cambridge, 1912; reprint: [A: Janiszewski, 1962; pp. 127–129].

[1913] *O rozcinaniu płaszczyzny przez continua* (On cutting up a plane with continua), Prace Matematyczno-Fizyczne 26, pp. 11–63; reprint: [A: Janiszewski, 1962; pp. 141–193].

[1962] *Oeuvres choisies*, Karol Borsuk, Jan Jaworowski, Bronisław Knaster, and Kazimierz Kuratowski (eds.), Warszawa: Państwowe Wydawnictwo Naukowe.

Mark KAC

[1936] *Sur les fonctions indépendantes I (Propriétés générales)*, Studia Math. 6, pp. 46–58.

[1938] *Sur les fonctions indépendantes V*, Studia Math. 7, pp. 96–100.

Mark KAC and Hugo STEINHAUS

[1936a] *Sur les fonctions indépendantes II (La loi exponentielle, la divergence de séries)*, Studia Math. 6, pp. 59–67; reprint: [A: Steinhaus, 1985; pp. 471–477].

[1936b] *Sur les fonctions indépendantes III (Le mouvement brownien, la loi de Maxwell)*, Studia Math. 6, pp. 89–97; reprint: [A: Steinhaus, 1985; pp. 478–485].

[1938] *Sur les fonctions indépendantes IV (Intervalle infini)*, Studia Math. 7, pp. 1–15; reprint: [A: Steinhaus, 1985; pp. 508–520].

Stefan KACZMARZ

[1930] *Über eine Anwendung der Funktionalen an die Orthogonalreihen*, Atti del Congresso Internazionale dei Matematici, Bologna, 3–10 Settembre 1928, Bologna: Nicolo Zanichelli, vol. III, pp. 189–191.

[1931] *Integrale von Dinischen Typus*, Studia Math. 3, pp. 189–191.

[1937] *Angenäherte Auflösung von Systemen linearer Gleichungen*, Bull. Intern. Acad. Polon. Sci. Lett., Cl. Sci. Math. Nat. A, pp. 355–357; English translation: *Approximate solution of systems of linear equations*, Intern. J. Control 57. 6 (1993), pp. 1269–1271.

Stefan KACZMARZ and Hugo STEINHAUS

[1936] *Theorie der Orthogonalreihen*, Monografie Matematyczne 6, Warszawa; translations: English (1951), Russian (1958).

Bronisław KNASTER, Kazimierz KURATOWSKI, and Stefan MAZURKIEWICZ

[1929] *Ein Beweis des Fixpunktsatzes für n-dimensionale Simplexe*, Fund. Math. 14, pp. 132–137; reprints: [A: Kuratowski, 1988; pp. 332–337] and [A: Mazurkiewicz, 1969; pp. 192–197].

Mirosław KRZYŻAŃSKI and Juliusz SCHAUDER

[1936] *Quasilineare Differentialgleichungen zweiter Ordnung von hyperbolischen Typus, Gemischte Randwertaufgaben*, Studia Math. 6, pp. 162–189; reprint: [A: Schauder, 1978; pp. 418–442].

Kazimierz KURATOWSKI

[1930] *Sur le problème de courbes gauches en topologie*, Fund. Math. 15, pp. 271–283; reprint: [A: Kuratowski, 1988; pp. 345–357]; English translation (translated by J. Jaworowski): [C: Borowiecki & Kennedy & Sysło, 1983; pp. 1–13].

[1931] *Evaluation de la classe borélienne ou projective d'un ensemble de points à l'aide des symboles logiques*, Fund. Math. 17, pp. 249–272; reprint: [A: Kuratowski, 1988; pp. 376–399].

[1933] *Topologie I*, Monografie Matematyczne 3, Warszawa-Lwów.

[1988] *Selected Papers*, Warszawa: PWN – Polish Scientific Publishers, pp. 345–357.

Kazimierz KURATOWSKI and Alfred TARSKI

[1931] *Les opérations logiques et les ensembles projectifs*, Fund. Math. 17, pp. 240–248; reprint: [C: Tarski, 1986; vol. IV, pp. 551–559]

Jean LERAY and Juliusz SCHAUDER

[1934] *Topologie et équations fonctionnelles*, Ann. de l'Ecole Norm. Supp. 51, pp. 45–78; reprint: [A: Schauder, 1978; pp. 320–348].

Antoni ŁOMNICKI

[1923] *Nouveaux fondements du calcul des probabilités*, Fund. Math. 4, pp. 34–71.

[1927] *Kartografja matematyczna* (Mathematical cartography), Lwów-Warszawa: Książnica-Atlas.

[1956] *Kartografia matematyczna* (Mathematical Cartography), second edition with changes, Warszawa: Państwowe Wydawnictwo Naukowe.

Zbigniew ŁOMNICKI and Stanisław ULAM

[1934] *Sur la théorie de la mesure dans les espaces combinatoires et son application au calcul des probabilités, I. Variables indépendantes*, Fund. Math. 23, pp. 237–278.

Stanisław MAZUR

[1928] *Über eine Limitierungsverfahren*, Math. Z. 28, pp. 599–611.

[1929] *Une remarque sur l'homéomorphie des champs fonctionnels*, Studia Math. 1, pp. 83–85.

[1930a] *Über die kleinste konvexe Menge, die eine gegebene kompakte Menge enthält*, Studia Math. 2, pp. 7–9.

[1930b] *Eine Anwendung der Theorie der Operationen bei der Untersuchung der Toeplitzschen Limitierungsverfahren, Erste Mitteilung*, Studia Math. 2, pp. 40–50.

[1933a] *Über konvexe Mengen in linearen normierten Räumen*, Studia Math. 4, pp. 70–84.

[1933b] *Über schwache Konvergenz in den Räumen (L^p)*, Studia Math. 4, pp. 128–133.

[1936] *O zbiorach i funkcjonałach wypukłych w przestrzeniach liniowych* (On convex sets and convex functionals in linear spaces)), Lwów, pp. 1–20.

[1938] *Sur les anneaux linéaires*, C. R. Mat. Acad. Sci. Paris 207, pp. 1025–1027.

Stanisław MAZUR and Władysław ORLICZ

[1933] *Über Folgen linearen Operationen*, Studia Math. 4, pp. 152–157; reprint: [A: Orlicz, 1988; vol. I, pp. 270–275].

[1934] *Grundlegende Eigenschaften der polynomischen Operationen*, Studia Math. 5, pp. 50–68 and 179–189; reprint: [A: Orlicz, 1988; vol. I, pp. 290–308].

[1948] *Sur les espaces métriques linéaires (I)*, Studia Math. 10, pp. 184–208; reprint: [A: Orlicz, 1988; vol. I, pp. 537–581].

[1953] *Sur les espaces métriques linéaires (II)*, Studia Math. 13, pp. 137–179; reprint: [A: Orlicz, 1988; vol. I, pp. 671–713].

Stanisław MAZUR and Juliusz SCHAUDER

[1936] *Über ein Prinzip in der Variationsrechnung*, Congrès Int. Math. Oslo 1.

Stanisław MAZUR and Stanisław ULAM

[1932] *Sur les transformations isométriques d'espaces vectoriels normés*, C. R. Math. Acad. Sci. Paris 194, pp. 946–948.

Stefan MAZURKIEWICZ

[1910] *Sur la théorie des ensembles*, C. R. Math. Acad. Sci. Paris 151, pp. 296–298; reprint: [A: Mazurkiewicz, 1969; pp. 27–29].

[1913] *O arytmetyzacji continuów* (On the arithmeticization of continua), C. R. de séances de la Société des Sciences et de Lettres de Varsovie, Cl. III, 6, pp. 305–311; *O arytmetyzacji continuów II*, *ibidem* 6, pp. 941–945; reprint: [A: Mazurkiewicz, 1969; pp. 37–45].

[1920] *Sur les lignes de Jordan*, Fund. Math. 1, pp. 166–209; reprint: [A: Mazurkiewicz, 1969; pp. 76–113].

[1969] *Travaux de topologie et ses applications*, Warszawa: Państwowe Wydawnictwo Naukowe.

Zbigniew MOROŃ

[1925] *O rozkładzie prostokątów na kwadraty* (On the decomposition of rectangles into squares), Przegląd Matematyczno-Fizyczny 3, pp. 151–152.

Władysław ORLICZ

[1929–1939] *Beiträge zur Theorie der Orthogonalentwicklungen I–VI*, Studia Math. 1, pp. 1–39 and 241–255; Bull. Intern. Acad. Polon. Sci., Sér. A, 8–9, pp. 229–238; Studia Math. 5, pp. 1–14; Studia Math. 6, pp. 20–38; Studia Math. 8, pp. 141–147; reprint: [A: Orlicz, 1988; vol. I, pp. 50–88, 89–103, 239–248, 276–289, 360–378, 403–409].

[1932] *Über eine gewisse Klasse von Räumen von Typus B*, Bull. Intern. Acad. Polon. Sci., Sér. A, 8/9, pp. 207–220; reprint: [A: Orlicz, 1988; vol. I, pp. 217–230].

[1933] *Über unbedingte Konvergenz in Funktionenräumen (I)*, Studia Math. 4, pp. 33–37; reprint: [A: Orlicz, 1988, vol. I; pp. 255–259].

[1936a] *Über Räume* (L^M), Bull. Intern. Acad. Polon. Sci. A, pp. 93–107; reprint: [A: Orlicz, 1988, vol. I; pp. 345–359].

[1936b] *Einige Gegenbeispiele zur Konvergenztheorie der allgemeinen Orthogonalentwicklungen*, Studia Math. 6, pp. 98–103; reprint: [A: Orlicz, 1988, vol. I; pp. 384–389].

[1988] *Collected Papers*, 2 vols., Warszawa: Państwowe Wydawnictwo Naukowe.

Józef PUZYNA

[1898–1900] *Teorya funkcyj analitycznych*, 2 vols., Lwów.

I. ROSENZWEIG and Hugo STEINHAUS

[1939] *Der Quadratleistungstarif*, Bulletin des Schweizerischen Elektrotechnischen Vereins 30, pp. 134–136.

Stanisław RUZIEWICZ

[1911] *Sur une classe des séries potentielles pantachiquement divergentes sur son circle de convergence*, Sprawozdania Towarzystwa Naukowego Warszawskiego 4, pp. 451–453.

[1913] *O funkcji ciągłej monotonicznej, nie posiadającej pochodnej w nieprzeliczalnej ilości punktów* (On continuous monotonic functions that are non-differentiable at uncountably many points), Sprawozdania Towarzystwa Naukowego Warszawskiego 6, pp. 282–305.

[1916] *O funkcyach ciągłych, monotonicznych, posiadających pantachiczne przedziały stałości* (On continuous monotonic functions having everywhere dense intervals of constancy), Prace Matematyczno-Fizyczne 27, pp. 19–31.

[1925] *Contributions à l'étude des ensembles de distances entre deux points*, Fund. Math. 7, pp. 141–143.

Juliusz Schauder

[1926] *The theory of surface measure*, Fund. Math. 8, pp. 1–48.

[1927] *Zur Theorie stetiger Abbildungen in Funktionalräumen*, Math. Z. 26, pp. 47–65 and 417–431; reprint: [A: Schauder, 1978; pp. 63–82].

[1929] *Invarianz des Gebietes in Funktionalräumen*, Studia Math. 1, pp. 123–139; reprint: [A: Schauder, 1978; pp. 147–162].

[1930a] *Über die Umkehrung linearer stetiger Funktionaloperationen*, Studia Math. 2, pp. 1–8; reprint: [A: Schauder, 1978; pp. 162–167].

[1930b] *Der Fixpunktsatz in Funktionalräumen*, Studia Math. 2, pp. 171–180; reprint: [A: Schauder, 1978; pp. 168–176].

[1932a] *Über den Zusammenhang zwischen der Eindeutigkeit und Lösbarkeit partieller Differentialgleichungen zweiter Ordnung vom elliptischen Typus*, Math. Ann. 106, pp. 661–721; reprint: [A: Schauder, 1978; pp. 235–297].

[1932b] *Sur le problème de Dirichlet généralisé pour les équations non linéaires du type elliptique*, C. R. Math. Acad. Sci. Paris 195, pp. 201–203; reprint: [A: Schauder, 1978; pp. 298–299].

[1933] *Über das Dirichletsche Problem im Großen für nicht-lineare elliptische Differentialgleichungen*, Math. Z. 37, pp. 623–634; reprint: [A: Schauder, 1978; pp. 305–316].

[1934a] *Sur les équations linéaires du type elliptique à coefficients continus*, C. R. Math. Acad. Sci. Paris 199, pp. 1366–1368; reprint: [A: Schauder, 1978; pp. 303–304].

[1934b] *Über lineare elliptische Differentialgleichungen zweiter Ordnung*, Math. Z. 38, pp. 257–282; reprint: [A: Schauder, 1978; pp. 354–379].

[1934c] *Sur les équations quasilinéares du type elliptique à coefficients continus*, C. R. Math. Acad. Sci. Paris 199, pp. 1566–1568; reprint [A: Schauder, 1978; pp. 352–353].

[1935a] *Das Anfangswertproblem einer quasilinearen hyperbolischen Differentialgleichung zweiter Ordnung in beliebiger Anzahl von unabhängigen Veränderlichen*, Fund. Math. 24, pp. 213–246; reprint: [A: Schauder, 1978; pp. 380–409].

[1935b] *Numerische Abschätzungen in elliptischen linearen Differentialgleichungen*, Studia Math. 5, pp. 34–42; reprint: [A: Schauder, 1978; pp. 410–417].

[1936a] *Gemischte Randwertaufgaben bei partiellen Differentialgleichungen vom hyperbolischen Typus*, Studia Math. 6, pp. 190–198; reprint: [A: Schauder, 1978; pp. 443–449].

[1936b] *Équations du type elliptique, problèmes linéaires*, Enseign. Math. 35, pp. 126–139; reprint: [A: Schauder, 1978; pp. 450–459].

[1936c] *Nichtlineare partielle Differentialgleichungen vom hyperbolischen Typus*, Congrès Int. Math. Oslo, pp. 60–61.

[1936d] *Einige Anwendungen der Topologie der Funktionalräume*, Matematičeskij Sbornik 1 (43), pp. 747–753; reprint: [A: Schauder, 1978; pp. 460–467].

[1937] *Cauchysches Problem für partielle Differentialgleichungen erster Ordnung, Anwendung einiger sich auf die Absolutbeträge der Lösungen beziehenden Abschätzungen*, Comment. Math. Helv. 9, pp. 263–287; reprint: [A: Schauder, 1978; pp. 468–487].

[1978] *Oeuvres*, Sous la rédaction de Jan Kisyński, Władysław Orlicz, Marceli Stark, Warszawa: PWN.

Wacław Sierpiński

[1908] *O pewnym twierdzeniu Cantora* (On a certain theorem of Cantor), Wiadom. Mat. 12, Suplement, pp. 31–37; French translation *Un théorème de Cantor* in: [Sierpiński, 1974–1976; vol. II, pp. 37–41].

[1910] *Nowy sposób dowodzenia twierdzenia Bolzano-Weierstrassa* (A new proof of the Bolzano-Weierstrass theorem), C. R. de séances de la Société des Sciences et de Lettres de Varsovie, Cl. III, 3, pp. 271 [French summary: *Nouvelle manière de démontrer le théorème de Bolzano-Weierstrass*].

[1912a] *Zarys teorii mnogości* (An outline of set theory), Biblioteka Matematyczno-Fizyczna, Warszawa (revised editions 1923, 1928).

[1912b] *O krzywych wypełniających kwadrat* (On square-filling curves), Prace Matematyczno-Fizyczne 23, pp. 193–219.

[1912c] *Sur une nouvelle courbe continue qui remplit toute une aire plane*, Bull. Intern. Acad. Sci. Cracovie A, pp. 462–478; reprint: [A: Sierpiński, 1974–1976; vol. II, pp. 52–66].

[1915] *Sur une courbe dont tout point est un point de ramification*, C. R. Math. Acad. Sci. Paris 160, pp. 302–305.

[1916a] *O krzywej, której każdy punkt jest punktem rozgałęzienia* (On a curve having all of its points as branch points), Prace Matematyczno-Fizyczne 27, pp. 77–85; French translation: *Sur une courbe dont tout point est un point de ramification* in: [A: Sierpiński, 1974–1976; vol. II, pp. 99–106].

[1916b] *Sur une courbe cantorienne qui contient une image biunivoque et continue de toute courbe donnée*, C. R. Math. Acad. Sci. Paris 162, pp. 629–632.

[1916c] *O kriwoj soderžaščej w sebe obraz wsiakoj kriwoj*, Matematičeskij Sbornik 30, pp. 267–287; French translation: *Sur une courbe cantorienne qui contient une image biunivoque et continue de toute courbe donnée*, in: [A: Sierpiński, 1974–1976; vol. II, pp. 107–119].

[1974–1976] *Oeuvres choisies*, 3 vols., Warszawa: Państwowe Wydawnictwo Naukowe.

Hugo Steinhaus

[1911] *Neue Anwendungen des Dirichlet'schen Prinzips*, Göttingen; reprint: [A: Steinhaus, 1985; pp. 47–87].

[1912] *O pewnym szeregu trygonometrycznym rozbieżnym*, Sprawozdania z posiedzeń Towarzystwa Naukowego Warszawskiego 5; English translation: *On a certain divergent trigonometric series*, in: [A: Steinhaus, 1985; pp. 109–112].

[1919] *Additive und stetige Funktionaloperationen*, Math. Z. 5, pp. 186–221; reprint: [Steinhaus, 1985; pp. 252–288].

[1920] *Sur les distances des points dans les ensembles de mesure positive*, Fund. Math. 1, pp. 93–103; reprint: [A: Steinhaus, 1985; pp. 296–304].

[1923a] *Les probabilités dénombrables et leur rapport à la théorie de mesure*, Fund. Math. 4, pp. 286–310; reprint: [A: Steinhaus, 1985; pp. 311–331].

[1923b] *Czem jest a czem nie jest matematyka?* (What is and what is not mathematics?), Lwów: H. Altenberg.

[1925] *Definicje potrzebne do teorii gier i pościgu* (Definitions for a theory of games and pursuit), Myśl Akademicka 1, pp. 13–14 (occasional publication) English translation: Naval Res. Logist. Quarter. 7 (1960), pp. 105–107.

[1927] *Czem jest matematyka i na czem polega jej postęp?*(What is mathematics and what does its progress depend on?), Kosmos 52.4, pp. 346–361; reprint: [B: Steinhaus, 2000; pp. 36–48].

[1929] *Anwendungen der Funktionalanalysis auf einige Fragen der reellen Funktionentheorie*, Studia Math. 1, pp. 51–81; reprint: [A: Steinhaus, 1985; pp. 386–413].

[1930a] *Sur la probabilité de la convergence de séries*, Studia Math. 2, pp. 21–39; reprint: [A: Steinhaus, 1985; pp. 435–451].

[1930b] *Quelques applications de l'analyse fonctionnelle à la théorie des fonctions d'une variable réelle*, Atti del Congresso Internazionale dei Matematici, Bologna, 3–10 Settembre 1928, Bologna: Nicolo Zanichelli, vol. III, pp. 183–187.

[1930c] *Über die Wahrscheinlichkeit dafür, dass der Konvergenzkreis einer Potenzreihe ihre natürliche Grenze ist*, Math. Z. 31, pp. 408–416; reprint: [A: Steinhaus, 1985; pp. 426–434].

[1938a] *La théorie et les applications des fonctions indépendantes au sens stochastique*, in: *Les fonctions aléatoires*, Colloque consacré à la théorie des probabilités, Paris: Herman, pp. 57–73; reprint: [A: Steinhaus, 1985; pp. 493–507].

[1938b] *Kalejdoskop matematyczny* (Mathematical Kaleidoscope), Lwów: Książnica-Atlas, p. 135; Polish revisions: PZWS 1954 and 1956, WSiP 1989; translations: English (*Mathematical Snapshots*, 1938, 1950, 1960, 1969), Russian (1949), Hungarian (1951), Czech (1953), Japanese (1957), German (1959), Romanian (1961), French (*Mathématiques instantanées*, 1964), Bulgarian (1974) and others.

[1940] *Sur les fonctions indépendantes VI, Equipartition*, Studia Math. 9, pp. 121–132; reprint: [A: Steinhaus, 1985; pp. 523–532].

[1985] *Selected Papers*, Warszawa: Państwowe Wydawnictwo Naukowe.

[2000] *Między duchem a materią pośredniczy matematyka* (Between spirit and matter there is mathematics), Warszawa-Wrocław: Wydawnictwo Naukowe PWN.

Stanisław ULAM

[1930] *Zur Masstheorie in der allgemeinen Mengenlehre*, Fund. Math. 16, pp. 140–150.

[1932] *Zum Massbegriffe in Produkträumen*, Verhandlungen des Internationalen Mathematiker-Kongress Zürich, vol. 2, pp. 118–119.

[1997] *O teorii miary w ogólnej teorii mnogości* (On measure theory in the general theory of sets) (reprint of his 1933 doctoral thesis), Wiadom. Mat. 33, pp. 155–168.

Eustachy ŻYLIŃSKI

[1925] *Some remarks concerning the theory of deduction*, Fund. Math. 7, pp. 203–209.

[1932] *Zur Begründung der Idealtheorie*, C. R. Soc. Sci. Varsovie 24, pp. 87–92.

B. Personal Recollections, Surveys, and Historical Source Material[111]

Albert, Z. (ed.) [1989], *Kaźń profesorów lwowskich, July 1941. Studia oraz relacje i dokumenty* (The execution of Lvov professors in 1941. Studies, accounts and documents), Wrocław: Wydawnictwo Uniwersytetu Wrocławskiego.

Albiński, M. [1976], *Wspomnienia o Banachu i Wilkoszu* (Remembering Banach and Wilkosz), Wiadom. Mat. 19.2, pp. 133–135.

Alexiewicz, A., Barański, F., and Koroński, J. [1997], *Koło Matematyczno-Fizyczne Studentów Uniwersytetu Jana Kazimierza we Lwowie* (The mathematics and physics student circle of Jan Kazimierz University in Lvov), in: X Szkoła Historii Matematyki, Zeszyty Naukowe Uniw. Opolskiego – Matematyka 30, pp. 9–42.

Axler, S. [1997], book review [B: Kałuża, 1996], Amer. Math. Monthly 104, p. 577.

Banach, S. jun. [1992], *O moim ojcu* (About my father), Delta 10.

Banach, S. [2001], *Autobiography* (reproduction of the Ukrainian original of 16 May 1945 and English translation), East J. Approx. 7.2, pp. 253–254; reproduction of the Ukrainian original [B: Bojarski & Ławrynowicz & Prytula, 2009; pp. 146].

Barański, F. [1993], *Lwowskie wspomnienia o Stefanie Banachu* (Memories of Stefan Banach in Lvov), Opuscula Math. 13, pp. 55–57.

Barone, J. and Novikoff, A. [1978], *A history of the axiomatic formulation of probability...*, Arch. Hist. Exact Sci. 18, pp. 123–190.

Birkhoff, G. and Kreyszig, E. [1984], *The establishment of Functional Analysis*, Hist. Math. 11, pp. 258–321.

Bojarski, B. [1980], *Przemówienie wygłoszone na uroczystości nadania stopnia doktora h.c. Prof. S. Mazurowi* (Oration delivered at the investiture ceremony for the honorary doctorate awarded to Prof. S. Mazur), Wiadom. Mat. 22.2, pp. 270–272.

Bojarski, B., Ławrynowicz, J., and Prytula, Ya. G. (eds.) [2009], *Lvov Mathematical School in the Period 1915–1945 as Seen Today*, Banach Center Publ. 87.

Bourbaki, N. [1980], *Elementy historii matematyki* (Elements of the history of mathematics), translated by S. Dobrzycki, Warszawa: Państwowe Wydawnictwo Naukowe.

Brzozowski, S. M. [1991], *Stanisław Ruziewicz*, Polski Słownik Biograficzny, tom 33.

[111]Not all the works listed in this part of the bibliography are cited explicitly in the main text of the book. However, they are all relevant and the author has made use of each and every one of them.

Chuyko, H. J. [2006], *Functional Analysis in Lviv after 1945*, in: [B: Bojarski & Ławrynowicz & Prytula, 2009; pp. 79–84].

Ciesielski, K. [1987], *Lost legends of Lvov, 1. The Scottish Café*, Math. Intelligencer 9.4, pp. 36–37.

Ciesielski, K. [1988], *Lost legends of Lvov, 2. Banach's Grave*, Math. Intelligencer 10.1, pp. 50–51.

Ciesielski, K. [1993], *O pewnych faktach związanych ze Stefanem Banachem*, Opuscula Math. 13, pp. 59–69. English summary: On some details of Stefan Banach's childhood, *ibidem*, pp. 71-74.

Ciesielski, K. and Pogoda, Z. [1988], *Conversation with Andrzej Turowicz*, Math. Intelligencer 10.4, pp. 13–20.

Czajkowskij, N. A. [1963], *Matiematika wo lwowskom uniwiersitietie s momienta jego wozniknowienija (1661) do 1918* (Mathematics at Lvov University from the year of its establishment (1661) to 1918), Woprosy istorii fizikomatiematiczeskich nauk, Moskwa.

Dauben, J. W. [1981], *The trigonometrical background to Georg Cantor's theory of sets*, Arch. Hist. Exact Sci. 7, pp. 181–216.

Dawidowicz, A. [1981], *Wspomnienia o Leonie Chwistku, Hugonie Steinhausie i Włodzimierzu Stożku* (Remembering Leon Chwistek, Hugo Steinhaus and Włodzimierz Stożek), Wiadom. Mat. 23.2, pp. 232–240.

Dawidowiczowa, A. [1989], *"Zeschnięte liście i kwiat...". Wspomnienia* ("Withered leaves and a flower" ... Recollections), Kraków: Wydawnictwo Literackie.

Derkowska, A. [1990], *Juliusz Paweł Schauder (1899–1943)*, in: S. Fudali (ed.), *Matematyka przełomu XIX i XX wieku* (Mathematics at the turn of the 19th and 20th centuries), Materiały z IV Ogólnopolskiej Szkoły Historii Matematyki, Uniwersytet Szczeciński: Materiały-Konferencje, Szczecin, pp. 39–43.

Derkowska, A., Mikosz, M., and Neugebauer, A. [1992], *Herman Auerbach (1901–1942)*, Wiadom. Mat. 29.2, pp. 225–231.

Dianni, J. and Wachułka, A. [1963], *Tysiąc lat polskiej myśli matematycznej* (A thousand years of Polish mathematical thought), Warszawa.

Dickstein, S. [1895], *Kilka słów o literaturze matematycznej polskiej w ciągu dwudziestolecia 1873–1892* (A few words about Polish mathematical literature between 1873–1892), Kosmos, zeszyt 8–10.

Dieudonné, J. [1981], *History of Functional Analysis*, Amsterdam: North-Holland.

Draus, J. [2007], *Uniwersytet Jana Kazimierza we Lwowie. Portret kresowej uczelni 1918–1946* (Jan Kazimierz University in Lvov. Portrait of a university in the borderlands 1918–1946). Kraków: Księgarnia Akademicka.

Drugi Zjazd [1931], *Drugi Zjazd Matematyków Polskich* (The Second Congress of Polish Mathematicians), Wiadom. Mat. 33, pp. 107–111.

Duda, R. [1995a], *"Fundamenta Mathematicae", "Studia Mathematica", "Acta Arithmeticae" – pierwsze trzy specjalistyczne czasopisma matematyczne* ("Fundamenta Mathematicae", "Studia Mathematica", "Acta Arithmeticae" – The first three specialist mathematical journals), Zeszyty Nauk. Politech. Śląskiej, Matematyka-Fizyka 76, pp. 47–80.

Duda, R. [1995b], *Życie matematyczne w Polsce w latach 1851–1950* (Mathematical life in Poland 1851-1950), in: S. Fudali (ed.) *Matematyka polska w stuleciu 1851–1950* (Polish mathematics over the century 1851–1950), Materiały z IX Ogólnopolskiej Szkoły Historii Matematyki, Uniwersytet Szczeciński: Materiały-Konferencje 16, Szczecin, pp. 7–24.

Duda, R. [1996], *"Fundamenta Mathematicae" and the Warsaw school of mathematics*, in: C. Goldstein, J. Gray, J. Ritter (eds.), *L'Europe mathématique Mythes, histoires, identités. Mathematical Europe – Myths, History, Identity*, Paris: Éditions de la Maison des sciences de l'homme, pp. 479–498.

Duda, R. [2001], *Sprawa akademika Łuzina* (The affair over the academic Łusin), Wiadom. Mat. 37, pp. 27–46.

Duda, R. [2004a], *Jeszcze o sprawie akademika Łuzina* (More about the affair over the academic Łuzin), Wiadom. Mat. 40, pp. 129–138.

Duda, R. [2004b], *On the Origins of Functional Analysis and the Lvov School*, Comm. Math., Thomus specialis in honorem Iuliani Musielak, pp. 5–45.

Duda, R. [2004c], *On the Warsaw interactions of logic and mathematics in the years 1919–1939*, Ann. Pure Appl. Logic 127, pp. 289–301.

Duda, R. [2005a], *The discovery of Banach spaces*, in: W. Więsław (ed.), *European Mathematics in the Last Centuries*, Conference held at Będlewo, April 26–30, 2004, Stefan Banach International Mathematical Center and Institute of Mathematics of Wrocław University, 2005, pp. 37–46.

Duda, R. [2005b], *Hugo Dyonizy Steinhaus*, Polski Słownik Biograficzny, tom 43/2, zeszyt 177.

Duda, R. [2008],*Od twierdzenia Baire'a o kategorii do przestrzeni polskich* (From Baire's category theorem to Polish spaces), Antiq. Math. 2, pp. 163–187.

Duda, R. [2009], *Osiągnięcia i znaczenie lwowskiej szkoły matematycznej* (The achievements and significance of the Lvov School of Mathematics), Prace Komisji Historii Nauki PAU, tom IX, pp. 35–52.

Duda, R. [2014], *Matematyka na uniwersytecie lwowskim* (Mathematics at Lvov University) [in press, Redzik].

Duda, R. [2011], *Matematyka na uniwersytecie lwowskim do usunięcia Polaków ze Lwowa* (Mathematics at Lvov University until the removal of Poles from Lvov), in: *Universitati Leopoliensi in memoriam*, Materiały sesji dla upamiętnienia trzysta pięćdziesiątej rocznicy jego fundacji, Kraków: PAU, pp. 291–323.

Dziwiński, P. [1890], *Rys działalności naukowej i nauczycielskiej Wawrzyńca Żmurki* (An outline of Wawrzyniec Żmurko's contributions to teaching and scholarship), Prace Matematyczno-Fizyczne 2, pp. 433–448.

Engelking, R. [1984], *O pracach Wacława Sierpińskiego z topologii* (On Wacław Sierpiński's works on topology), Wiadom. Mat. 26.1, pp. 18–24.

Estreicher, K. [1979], *Leon Chwistek. Biografia artysty* (Leon Chwistek. A biography of the artist), Warszawa.

Felgner, U. [2002], *Der Begriff der Funktion*, in: F. Hausdorff, Gesammelte Werke, Band II, Berlin: Springer, pp. 621–633.

Finkel, L. and Starzyński, S. [1894], *Historya Uniwersytetu Lwowskiego*, Lwów.

Fras, Z. [1999], *Galicja, Wrocław*: Wydawnictwo Dolnośląskie .

Girlich, H.-J. [2005], *Łomnicki-Steinhaus-Kolmogorov: Steps to a modern probability theory*, in: Więsław, W. (ed.), *European Mathematics in the Last Centuries*, Conference held at Będlewo, April 26–30, 2004, Stefan Banach International

Mathematical Center and Institute of Mathematics of Wrocław University, pp. 47–56.

Gołąb, S. [1947], *O dorobku matematyków polskich w nauce światowej* (The contributions of Polish mathematicians to world knowledge), Życie Nauki, pp. 13–14.

Granas, A. [1981], *KKM-maps and their applications to non-linear problems*, in: [B: Mauldin, 1981; pp. 45–61].

Guillaume, M. [1994], *La logique mathématique en sa jeunesse*, in: [B: Pier, 1994], pp. 185–367.

Hochstadt, H. [1980], *E. Helly. Father of the Hahn-Banach Theorem*, Math. Intelligencer 2.

Holgate, P. [1997], *Independent functions: Probability and analysis in Poland between the wars*, Biometrika 84, pp. 161–173.

Hollanek, E. and Hollanek, A. [1990], *I zobaczyć miasto Lwów* (A look at the city of Lvov), Rzeszów: KAW.

Hryciuk, G. [2000], *Polacy we Lwowie 1939–1944* (The Poles in Lvov 1939–1944). Warszawa: Życie codzienne.

Hurwic, J. (ed.) [1967], *Wkład Polaków do nauki. Nauki ścisłe* (The scholarly contributions of Poles. The exact sciences), Biblioteka Problemów 101, Warszawa: Państwowe Wydawnictwo Naukowe.

Iwiński, T. [1975], *Ponad pół wieku działalności matematyków polskich. Zarys działalności Polskiego Towarzystwa Matematycznego 1919–1973* (More than a century of work of Polish mathematicians. An outline description of the work of the Polish Mathematical Society 1919–1973.), Warszawa: Państwowe Wydawnictwo Naukowe.

Jach, D. [1990], *Mark Kac (1914–1984)*, in: S. Fudali (ed.), *Matematyka przełomu XIX i XX wieku* (Mathematics at the turn of the 19th and 20th centuries), Materiały z IV Ogólnopolskiej Szkoły Historii Matematyki, Uniwersytet Szczeciński: Materiały-Konferencje, Szczecin, pp. 131–133.

Jahn, A. [1991], *Z Kleparowa w świat szeroki* (Away from Kleparow into the big wide world), Wrocław: Wydawnictwo Ossolineum.

Jakimowicz, E. and Miranowicz, A. (eds.) [2011], *Stefan Banach: Remarkable Life, Brilliant Mathematics*, third ed., American Math. Soc., 2011.

James, I. [2002], *Remarkable Mathematicians from Euler to von Neumann*, Cambridge University Press.

Janas, J. [1993], *Twierdzenie spektralne w aspekcie historycznym* (Spectral theorem from a historical perspective), Opuscula Math. 13, pp. 37–43.

Janiszewski, Z. [1917], *Stan i potrzeby matematyki w Polsce* (The current state and needs of mathematics in Poland), in: *Nauka polska, jej potrzeby, organizacja i rozwój* (Learning in Poland, its needs, its organization and its development), Warszawa, pp. 11–18; reprint: Wiadom. Mat. 7 (1963), pp. 3–8.

Jaworski, F. [1912], *Uniwersytet Lwowski. Wspomnienia jubileuszowe* (Lvov University. Jubilatory memoirs), Lwów: Towarzystwo Miłośników Przeszłości Lwowa.

Kac, M. [1974], *Hugo Steinhaus – a reminiscence and tribute*, Amer. Math. Monthly 81, pp. 572–581.

Kac, M. [1978], *Henri Lebesgue i polska szkoła matematyczna: obserwacje i wspomnienia* (Henri Lebesgue and the Polish mathematical school: observations and recollections), Wiadom. Mat. 20, pp. 189–192.

Kac, M., Hartman, S., Schinzel, A., Żelazko, W., Kahane, J.-P., and Semadeni, Z. [1979], *Wspomnienia o Marcelim Starku* (Remembering Marceli Stark), Wiadom. Mat. 21.2, pp. 104–113.

Kac, M. [1985], *Enigmas of Chance. An Autobiography*, New York; Polish translation: *Zagadki losu*, translated by K. and S. Lipszycowie, Polska Fundacja Upowszechniania Nauki, Warszawa, 1997.

Kahane, J.-P. [1992], *Aperçu sur l'influence de l'école mathématique polonaise 1918-1939*, Centre Scientifique de l'Academie polonaise de Sciences Paris. A Polish translation appeared in Wiadom. Mat. 31, 1995, pp. 163–175.

Kalbarczyk, S. [2001], *Polscy pracownicy nauki. Ofiary zbrodni sowieckich w latach II wojny światowej. Zamordowani, więzieni, deportowani* (Polish employees working in the academic sector. The victims of Soviet crimes during World War II. The murdered, the imprisoned and the deported.), Warszawa: Wydawnictwo Neriton.

Kałuża, R. [1992], *Stefan Banach*, Warszawa: Wydawnictwo GZ.

Kałuża, R. [1996], *The Life of Stefan Banach*, translated and edited by A. Kostant and W. Woyczyński, Boston-Basel-Berlin: Birkhäuser.

Kasa [1992], *Kasa imienia Józefa Mianowskiego. Fundacja popierania nauki 1881–1991* (The Józef Mianowski Fund. A fund for supporting scholarship 1881-1991), Warszawa, 1992.

Kąkol, J. and Palka, Z. (eds.) [2002], *Władysław Orlicz (1903–1990) – Twórca Poznańskiej Szkoły Matematycznej* (Władysław Orlicz – the founder of the Poznań School of Mathematics), Poznań: Wydawnictwo Naukowe UAM.

Kennedy, J. W., Quintas, L. V., and Sysło, M. M. [1985], *The theorem on planar graphs*, Hist. Math. 12, pp. 356–368.

Köthe, G. [1994], *Wkład Stanisława Mazura w analizę funkcjonalną* (Stanisław Mazur's contribution to functional analysis), Wiadom. Mat. 30.2, pp. 199–250.

Kowarzyk, H. [1973], *Współpraca Hugona Steinhausa z medycyną i medykami* e (Hugo Steinhaus's collaborative work with medical staff and on medicine), Wiadom. Mat. 17, pp. 65–69.

Kozielecki, J. [1999], *Banach – geniusz ze Lwowa* (Banach – the genius from Lvov), Warszawa: Wydawnictwo Akademickie "Żak".

Krawczyński, W. [2009], *Przez tundrę i tajgę po sowieckich łagrach* (Through tundra and tajga after Soviet gulags), Wrocław: Biblioteka Zesłańca, Polskie Towarzystwo Ludoznawcze.

Księga Pamiątkowa [1929], *Księga Pamiątkowa I Polskiego Zjazdu Matematycznego, Lwów 7–10.IX.1927* (Commemorative publication for the 1st Polish Mathematical Congress, Lvov, 7–10.IX.1927), Appendix to "Annales de la Société Polonaise de Mathématique", Kraków.

Kuratowski, K. [1969], *Polskie Towarzystwo Matematyczne w okresie międzywojennym* (The Polish Mathematical Society during the inter-war years), Nauka Polska 17.6 (84), pp. 65–69.

Kuratowski, K. [1973], *Pół wieku matematyki polskiej 1920–1970* (A half century of Polish mathematics 1920–1970), Biblioteka Wiedzy Powszechnej Omega 247, Warszawa: Wiedza Powszechna.

Kuratowski, K. [1980], *A Half Century of Polish Mathematics. Remembrances and Recollections*, Warsaw: Oxford Pergamon Press and Polish Scientific Publishers.

Kuratowski, K. [1981], *Notatki do autobiografii* (Autobiographical notes), Warszawa: Czytelnik.

Kuratowski, K. [1989], *The Past and Present of the Polish School of Mathematics*, in: I. Stasiewicz-Jasiukowa (ed.), *The Founders of Polish Schools and Scientific Models Write about Their Works*, Wrocław-Warszawa: Ossolineum and The Polish Academy of Sciences Press, pp. 47–80.

Kwapień, S. [1990], *Stanisław Mazur. Życie i działalność naukowa* (Stanisław Mazur. His life The life and academic work.), in: S. Fudali (ed.), *Matematyka przełomu XIX i XX wieku* (Mathematics at the turn of the 19th and 20th centuries), Materiały z IV Ogólnopolskiej Szkoły Historii Matematyki, Uniwersytet Szczeciński: Materiały-Konferencje, Szczecin, pp. 57–68.

Lance, W., Pliczko, A., and Storoż, O. [1992], *Ob issledowanijach lwowskich matiematikow po funkcyonalnomu analizu posle 1945 goda*, Lwów [preprint].

Lanckorońska, K. [2008], *Wspomnienia wojenne* (War Memoirs) *22.IX.1939–5.IV.1945*, Słowo wstępne L. Kalinowski and E. Orman, Kraków: Wyd. Znak.

Leja, F. [1969], *Powstanie Polskiego Towarzystwa Matematycznego* (The foundation of the Polish Mathematical Society), Wiadom. Mat. 12.1, pp. 3–8.

Leopolis docet [2001], arcade exhibition in Wrocław, in memory of the Lvov Professors murdered in 1941.

Leray, J. [1959], *O twórczości J. Schaudera* (On the creative output of J. Schauder), Wiadom. Mat. 3.1, pp. 11–19.

Leray, J. [1980], *O moim przyjacielu Juliuszu Schauderze* (About my friend Juliusz Schauder), Wiadom. Mat. 23.1, pp. 75–83.

Lewickyj, B. [1960], *Sprawa dr Oberlaendera* (The case of Dr Oberlaender), Kultura 1960, no. 1–2; see also: B. Łewicki, *Terror i rewolucja*, Warszawa: LTW (n.d.), p. 12.

Łomnicki, A. and Ruziewicz, S. [1921], *Józef Puzyna (1856–1919)*, Wiadom. Mat. 25, pp. 113–119.

Łukaszewicz, J. [1973], *Rola Hugona Steinhausa w rozwoju zastosowań matematyki* (The role of Hugo Steinhaus in the advancement of applied mathematics), Wiadom. Mat. 17, pp. 39–50.

Maligranda, L. [2002], *Władysław Orlicz (1903–1990) – jego życie i wkład do matematyki* (Wladysław Orlicz (1903-1990) – his life and contribution to mathematics), in: [B: Orlicz, 2002; pp. 33–80].

Maligranda, L. [2004], *Setna rocznica urodzin Władysława Orlicza* (Centenary anniversary of the birth of Władysław Orlicz), in: W. Więsław (ed.), *Matematyka abelowa – w dwóchsetlecie urodzin Nielsa Henrika Abela (1802–1829)* (Abelian Mathematics – on the bicentenary of the birth of Niels Henrik Abel (1802-1829)), XVII Ogólnopolska Szkoła Historii Matematyki, Nowy Sącz, 9–13 June 2003, Nowy Sącz: Państwowa Wyższa Szkoła Zawodowa, pp. 73–126.

Maligranda, L. [2007], *Stefan Kaczmarz (1895–1939)*, Antiq. Math. 1, pp. 15–61.

Maligranda, L. [2008], *Antoni Łomnicki*, Wiadom. Mat. 44, pp. 61–112.

Maligranda, L. and Wnuk, W. [2000], *Władysław Orlicz (1903–1990)*, Wiadom. Mat. 36, pp. 85–134.

Marciniak, A. [2002], *Tradycje Lwowskiej Szkoły Matematycznej w badaniach Profesora Władysława Orlicza (1903–1990)* (The tradition of the Lvov School of Mathematics in the research of Professor Władysław Orlicz (1903–1990)), Biuletyn Archiwum PAN, Warszawa, no. 43, pp. 140–151.

Marczewski, E. [1948], *Sur l'œuvre scientifique de Stefan Banach*, II. Théorie des fonctions réelles et théorie de la mesure, Colloq. Math. 1, pp. 93–102.

Marczewski, E. [1962], *Zygmunt Janiszewski*, Polski Słownik Biograficzny, tom 10, pp. 527–529.

Marczewski, E. [1967], *Hugo Steinhaus*, Nauka Polska 15, pp. 82–97.

Marczewski, E. [1973], *Steinhaus*, Wiadom. Mat. 17, pp. 101–108.

Marczewski, E. [1985], *Hugo Steinhaus*, in: [A: Steinhaus, 1985], pp. 11–24.

Matuszewska, W. [1968], *Władysław Orlicz*, Nauka Polska 16.5.

Mauldin, R. D. [1981], *The Scottish Book. Mathematics from the Scottish Café*, Boston-Basel-Stuttgart: Birkhäuser.

Mazur, S. [1961], *Przemówienie wygłoszone na uroczystości ku uczczeniu pamięci Stefana Banacha* (Speech delivered in commemoration of Stefan Banach), Wiadom. Mat. 4.3, pp. 249–250.

Mazur, S. [1980], *Przemówienie wygłoszone przy nadaniu doktoratu honorowego Uniwersytetu Warszawskiego* (Speech delivered at the award from Warsaw University of his honorary doctorate), Wiadom. Mat. 22.2, pp. 266–272.

Mierzecki, R. [2009], *Budowa wrocławskiego pomnika w latach 1956–1964 ku czci polskich profesorów zamordowanych we Lwowie w 1941 r.* (The construction between 1956–1964 of the Wrocław cenotaph honoring the Polish professors murdered in Lvov in 1941.), Analecta 18.1–2, pp. 7–50.

Mioduszewski, J. [2004], *Stanisław Ruziewicz (1889–1941) po przeszło stu latach*, in: W. Więsław (ed.), *Matematyka abelowa – w dwóchsetlecie urodzin Nielsa Henrika Abela (1802–1829)* (Abelian Mathematics – on the occasion of the bicentenary of the birth of Niels Henrik Abel (1802-1829)), XVII Ogólnopolska Szkoła Historii Matematyki, Nowy Sącz, 9–13 June 2003, Nowy Sącz: Państwowa Wyższa Szkoła Zawodowa, pp. 127–136.

Mitropolskij, Yu. A. and Breus, K. A. [1967], *Osnownyje issledowanja Instituta Matematiki USSR za gody sowietskoj własti*, Ukr. Mat. Ż. 19.6, pp. 16–31.

Monna, A. F. [1975], *Banach's Théorie des opérateurs linéaires*, Nieuw Arch. Wisk. 23, pp. 67–71.

Moore, G. H. [1982], *Zermelo's Axiom of Choice, Its Origins, Development and Influence*, Studies in the History of Mathematics and Physical Sciences 8, New York-Heidelberg-Berlin: Springer.

Murawski, R. [1995], *Filozofia matematyki. Zarys dziejów* (The philosophy of mathematics. An outline of works.), Warszawa: Wydawnictwo Naukowe PWN.

Musielak, J. and Wnuk, W. [1996], *Andrzej Alexiewicz (1917–1995)*, Wiadom. Mat. 32, pp. 153–163.

Nicieja, S. S. [2001], *Martyrologia profesorów lwowskich w lipcu 1941 roku* (The murder of Lvov professors in July 1941), in: [B: Leopolis docet, 2001].

Nikolski, S. M. [1993], *Wspomnienie o Stefanie Banachu* (Remembering Stefan Banach), Wiadom. Mat. 30.1, pp. 115–120.

Obituarium [1982], *Mazur, S., 1. January 1905 – 5. November 1982*, Studia Math. 71, pp. 223–224.

Olech, C. [1975], *Matematyka wyprzedza praktykę* (Mathematics overtakes practice), in: *Nauka polska i jej osiągnięcia* (Polish science and its achievements), Biblioteka Problemów 202, Warszawa: Państwowe Wydawnictwo Naukowe, pp. 11–61.

Olesiak, Z. [2001], *O lwowskim środowisku matematycznym podczas wojny* (On the mathematical culture in Lvov during the war), Kwart. Historii Nauki i Techniki 46, pp. 35–59.

Olesiak, Z. and Sułym, O. [2001], *Władysław Nikliborc*, Kwart. Historii Nauki i Techniki 46.3, pp. 113–124.

Olszewicz, B. [1947], *Lista strat kultury polskiej (1 IX 1939–1 III 1946)* (List of losses to Polish culture (1 IX 1939–1 III 1946)), Warszawa: S. Aret.

Opial, Z. [1966], *Dzieje nauk matematycznych w Polsce* (History of the mathematical sciences in Poland), Studia i Materiały z Dziejów Nauki Polskiej, Seria B, tom 10, pp. 137–166.

Orlicz, W. [1948], *Sur l'oeuvre scientifique de Stefan Banach, I. Théorie des opérations et théorie des séries orthogonales*, Colloq. Math. 1, pp. 81-92.

Orlicz, W. [1965a], *Stanisław Mazur*, Nauka Polska 13.1.

Orlicz, W. [1965b], *Referat o działalności naukowej prof. H. Steinhausa wygłoszony przy nadaniu doktoratu honorowego przez Uniwersytet im. Adama Mickiewicza w Poznaniu* (Lecture on the scholarly contributions of Prof. H. Steinhaus, delivered upon the award of an honorary doctorate from Adam Mickiewicz University in Poznań), Wiadom. Mat. 8, pp. 109–118.

Orlicz, W. [1973], *Antoni Łomnicki*, Polski Słownik Biograficzny, tom 18, pp. 388–389.

Orlicz, W. [1980], *Przemówienie wygłoszone przy nadaniu doktoratu honorowego Politechniki Poznańskiej* (Speech delivered upon the award of an honorary doctorate from Poznań Polytechnic), Wiadom. Mat. 22, pp. 279–284.

Orlicz, W. [1981], *Lwowska Szkoła Matematyczna w okresie międzywojennym* (The Lvov School of Mathematics during the inter-war years), Wiadom. Mat. 23, pp. 222–231.

Orlicz, W. [1985], *Stefan Kaczmarz (1895–1939)*, Wiadom. Mat. 26.2, 155–164.

Orlicz, W. [1988], *Achievements of Polish mathematicians in the domain of functional analysis in the years 1919–1951*, Nauka Polska 4; reprint: [A: Orlicz, 1988; tom 2, pp. 1616–1641].

Parks, P. C. [1993], *Stefan Kaczmarz (1895–1939)*, Intern. J. Control 57.6, pp. 1263–1267.

Pawlikowska-Brożek, Z. [1990], *Stefan Banach w świetle wspomnień* (A retrospective recollection of Stefan Banach), in: S. Fudali (ed.), *Matematyka przełomu XIX i XX wieku* (Mathematics at the turn of the 19th and 20th centuries), Materiały IV Ogólnopolskiej Szkoły Historii Matematyki, Uniwersytet Szczeciński: Materiały-Konferencje, Szczecin, pp. 101–112.

Pawlikowska-Brożek, Z. [1992], *Wykaz profesorów i docentów matematyki pracujących w polskich uczelniach w latach 1919–1939* (List of professors and readers in mathematics employed at Polish places of learning between 1919–1939), Wiadom. Mat. 24.2, pp. 219–223.

Pawlikowska-Brożek, Z. [1995], *Zbigniew Łomnicki (1904–1994)*, Wiadom. Mat. 31, pp. 187–190.

Pełczyński, A. and Semadeni, Z. [1969], *Rozwój analizy funkcjonalnej w Polsce* (The development of functional analysis in Poland), Wiadom. Mat. 12.1, pp. 83–108.

Pier, J.-P. (ed.) [1994], *Development of Mathematics 1900–1950*, Basel-Boston-Berlin: Birkhäuser.

Płoski, A. [1988], *O dziele Józefa Puzyny "Teorya funkcji analitycznych"* (About Józef Puzyna's opus "The theory of analytic functions"), in: S. Fudali (ed.), *Matematyka XIX wieku* (Mathematics during the 19th century), Materiały II Ogólnopolskiej Szkoły Historii Matematyki, Uniwersytet Szczeciński: Materiały-Konferencje, Szczecin, pp. 237–244.

Podraza, A. [1992], *Uniwersytet Lwowski w latach II wojny światowej* (Lvov University during World War II), in: [B: Stasiewicz-Jasiukowa, 1992; pp. 34–51].

Politechnika [1932], *Politechnika Lwowska, jej stan obecny i potrzeby* (Lvov Polytechnic, its needs and current status), Lwów. Politechnika Lwowska [1993], Politechnika Lwowska, praca zbiorowa, Wrocław: Wydawnictwo Politechniki Wrocławskiej.

Popławski, Z. [1992], *Dzieje Politechniki Lwowskiej 1844–1945* (History of Lvov Polytechnic 1844-1945), Wrocław-Warszawa-Kraków: Wydawnictwo Ossolineum.

Prytula, Ya. G. [2009], *Remarks on the history of mathematics in Lvov up to the middle of the XXth century*, in: [B: Bojarski & Ławrynowicz & Prytula, 2009], pp. 17–26.

Przeniosło, M. [2007], *Twórcy lwowskiej szkoły matematycznej* (The founders of the Lvov School of Mathematics), Dzieje Najnowsze, z. 2, pp. 59–76.

Puzyna, J. [1889], *Profesor Wawrzyniec Żmurko, jego życie i dzieło* (Professor Wawrzyniec Żmurko, his life and works), Kosmos 14, pp. 168–184.

Redzik, A. [2006], *Wydział Prawa Uniwersytetu Lwowskiego w latach 1939–1946* (The Law Department at Lvov University during 1939-1946), Lublin: Towarzystwo Naukowe KUL.

Redzik, A. [2007], *Szkic o dziejach uniwersytetu lwowskiego w latach 1939–1946* (Outline sketch of the history of the university in Lvov between 1939-1946), Visnyk Lviv Univ., Ser. histor., Special issue, pp. 577–592.

Rota, G.-C. [1990], *Utracona kawiarnia (wspomnienia o Stanisławie Ulamie)*, Preprint Inst. Mat. PAN, Seria B, no. 24, July 1990. See also *The Lost Café* (*ibidem*), Los Alamos Science Special Issue 1987, pp. 23–32.

Ryll-Nardzewski, C. [1973], *Prace Hugona Steinhausa o sytuacjach konfliktowych* (The works of Hugo Steinhaus on situations of conflict), Wiadom. Mat. 17, pp. 29–38.

SBMP [2003], S. Domaradzki, Z. Pawlikowska-Brożek, and D. Węglowska (eds.), *Słownik Biograficzny Matematyków Polskich* (Biographical Dictionary of Polish Mathematicians), Tarnobrzeg.

Schaerf, H. M. [1982], *Wspomnienie o Juliuszu Schauderze* (Remembering Juliusz Schauder), Wiadom. Mat. 24.1, pp. 91-94.

Schinzel, A. [1984], *Rola Wacława Sierpińskiego w historii matematyki polskiej* (The role of Wacław Sierpiński in the history of Polish mathematics), Wiadom. Mat. 26.1, pp. 1-9.

Shafer, G. and Vovk, V. [2006], *The sources of Kolmogorov's Grundbegriffe*, Statistical Science 21, pp. 70-98.

Shields, A. [1989], *Banach Algebras, 1939–1989*, Math. Intelligencer 113, pp. 15–17.

Siegmund-Schulze, R. [1892], *Die Anfänge der Funktionalanalysis und ihr Platz im Umwälzungsprozess der Mathematik um 1900*, Arch. Hist. Exact Sci. 26, pp. 13–71.

Sierpiński, W. [1947], *Matematyka polska w czasie wojny i po wojnie* (Polish mathematics during and after the war), Nauka Polska 25, pp. 90-97.

Sierpiński, W. [1958], *Matematyka w Polsce* (Mathematics in Poland), Życie Szkoły Wyższej 6, pp. 1–10.

Sierpiński, W. [1963], *O polskiej szkole matematycznej* (On the Polish School of Mathematics), Problemy 3 (204), pp. 146–155.

Sierpiński, W. [1967], *O polskiej szkole matematycznej* (On the Polish School of Mathematics), in: [B: Hurwic, 1967], pp. 413–434.

Ślebodziński, W. [1948], *Władysław Nikliborc et son oeuvre scientifique*, Colloq. Math. 1, pp. 322–330.

Ślebodziński, W. [1969], *Kazimierz Żorawski*, Wiadom. Mat. 11.1, pp. 49–64.

Snitko-Rzeszut, J. [1989], *Stefan Marian Kaczmarz*, Wojskowy Przegląd Historyczny 34.4, p. 289.

Sobolew, S. Ł. [1961], *Przemówienie wygłoszone na uroczystości ku uczczeniu pamięci Stefana Banacha* (Speech delivered in a commemoration ceremony for Stefan Banach), Wiadom. Mat. 4.3, pp. 261–264.

Stark, M. [1973], *Hugo Steinhaus jako nauczyciel w okresie lwowskim* (Hugo Steinhaus as a teacher based in Lvov), Wiadom. Mat. 17, pp. 77–84.

Stasiewicz-Jasiukowa, I. (ed.) [1992], *Lwowskie środowisko naukowe w latach 1939–1945* (The academic scene in Lvov during 1939–1945), wyd. III, Warszawa: Polska Akademia Nauk.

Steen, L. A. [1976], *Momenty zwrotne w historii teorii spektralnej* (Breakthrough moments in the history of spectral theory), Wiadom. Mat. 19.2, pp. 141–164.

Steinhaus, H. [1948], *Souvenirs de Stefan Banach*, Colloq. Math. 1, pp. 74–80.

Steinhaus, H. [1960], *Stefan Banach (1892–1945)*, Nauka Polska,
reprints: Wiadom. Mat. 4.3 (1961), pp. 252–259; [B: Hurwic, 1967; pp. 435–448].

Steinhaus, H. [1961a], *Stefan Banach*, Wiadom. Mat. 4.3, pp. 251–259.

Steinhaus, H. [1961b], *Stefan Banach*, 1896–1945, Scripta Math. 26, pp. 93–100.

Steinhaus, H. [1965], Przemówienie wygłoszone przy nadaniu doktoratu honorowego przez Uniwersytet im. Adama Mickiewicza w Poznaniu (Speech delivered upon the award of his honorary doctorate from Adam Mickiewicz University in Poznań), Wiadom. Mat. 8, pp. 119–125.

Steinhaus, H. [1973], *Autobiografia*, Wiadom. Mat. 17, pp. 3–11.

Steinhaus, H. [1974], *Zadaczi i razmyszlenija*, Mir [a selection of texts in Russian].

Steinhaus, H. [1992], *Słownik racjonalny* (A rational dictionary), Wrocław-Warszawa-Kraków: Wydawnictwo Ossolineum.

Steinhaus, H. [2010], *Wspomnienia i zapiski* (Notes and recollections) third edition, with corrections, Centrum Steinhausa, Politechniki Wrocławskiej, Wrocław: Atut.

Stone, M. H. [1961], *Nasz dług wobec Stefana Banacha* (Our debt to Stefan Banach), Wiadom. Mat. 4.3, pp. 265–267.

Straszewicz, S. [1963], *Ze wspomnień o Zygmuncie Janiszewskim* (Recollections about Zygmunt Janiszewski), Wiadom. Mat. 8, pp. 131–133.

Studnicki, G. [1985], *Przyczynek do biografii Władysława Nikliborca (list do Redakcji)* (Contribution to the biography of Władysław Nikliborc (Letter to the Editor)), Wiadom. Mat. 26, pp. 231–232.

Szałajko, K. [1984], *Wspomnienia o Kole Matematyczno-Fizycznym Studentów Uniwersytetu Jana Kazimierza we Lwowie* (Memories of the Mathematics and Physics Student Circle at Jan Kazimierz University in Lvov), Wiadom. Mat. 26.1, pp. 85-96.

Szałajko, K. [1990], *Antoni Łomnicki (1881–1941)*, in: S. Fudali (ed.), *Matematyka przełomu XIX i XX wieku* (Mathematics at the turn of the 19th and 20th centuries), Materiały IV Ogólnopolskiej Szkoły Historii Matematyki, Uniwersytet Szczeciński: Materiały-Konferencje, Szczecin, pp. 113–122.

Szałajko, K. [1993], *Wspomnienia o Stefanie Banachu na tle Lwowa i lwowskiej szkoły matematycznej* (Recollections of Stefan Banach under the backdrop of Lvov and the Lvov School of Mathematics), Opuscula Math. 13, pp. 45–54.

Szałajko, K. [1994], *Wspomnienia lwowskie* (Memories of Lvov), Wiadom. Mat. 30.2, pp. 251–263.

Szökefalvi-Nagy, B. [1961], *Przemówienie na uroczystości ku uczczeniu pamięci Stefana Banacha* (Memorial speech in commemoration of Stefan Banach), Wiadom. Mat. 4.3, pp. 265–268.

Towarzystwo [1932], *Towarzystwo Naukowe Warszawskie* (The Warsaw Scientific Society), Warszawa.

Trzeci Zjazd [1937], *IIIe Congrès Polonais de Mathématiques*, Ann. Soc. Polon. Math. 16, pp. 181–212.

Trzeci Zjazd [1939], *Trzeci Zjazd Matematyków Polskich (28 IX–3 X 1937)* (The Third Congress of Polish Mathematicians (28 Sept.–3 Oct. 1937)), Wiadom. Mat., pp. 149–157.

Trzynadlowski, J. [1973], *Sztuka słowa Hugona Steinhausa* (Hugo Steinhaus's word art), Wiadom. Mat. 17, pp. 71–76.

Tucholski, J. [1991], *Mord w Katyniu. Kozielsk, Ostaszków, Starobielsk. Lista ofiar* (Murder in Katyn, Kozielsk, Ostaszków and Starobielsk. A list of victims), Warszawa: Pax.

Turowicz, A. [1973], *Wspomnienia o profesorze Steinhausie* (Remembering Professor Steinhaus), Wiadom. Mat. 17, pp. 85–89.

Turowicz, A. [1979], *Wspomnienie o przyjacielu [M. Starku]* (Remembering a friend [M. Stark]), Wiadom. Mat. 21.1, pp. 102–104.

Turowicz, A. [1990], *Fragment wspomnień o Stanisławie Mazurze* (Some recollections about Stanisław Mazur), in: S. Fudali (ed.),*Matematyka przełomu XIX i XX wieku* (Mathematics at the turn of the 19th and 20th centuries), Materiały IV Ogólnopolskiej Szkoły Historii Matematyki, Uniwersytet Szczeciński: Materiały-Konferencje, Szczecin, pp. 69–70.

Turowicz, A. [1995], *W sprawie dowodu twierdzenia Weierstrassa-Stone'a* (About the proof of the Weierstrass-Stone theorem), Wiadom. Mat. 31, pp. 149–150.

Ulam, S. [1946], *Stefan Banach (1892–1945)*, Bull. Amer. Math. Soc. 52, pp. 600–603.

Ulam, S. [1969], *Wspomnienia z Kawiarni Szkockiej* (Recollections of the Scottish Café), Wiadom. Mat. 12.1, pp. 49–58.

Ulam, S. [1976], *Adventures of a Mathematician*, 1976; Polish translation: *Przygody matematyka*, translated by A. Górnicka, Warszawa: Prószyński i S-ka 1996.

Uniwersytet [1938], *Uniwersytet Jana Kazimierza we Lwowie* (Jan Kazimierz University in Lvov), Skład Uniwersytetu w roku akademickim 1938/1939, Lwów.

Urbanik, K. [1973], *Idee Hugona Steinhausa w teorii prawdopodobieństwa* (Hugo Steinhaus's ideas in probability theory), Wiadom. Mat. 17, pp. 39–50.

Wachułka, A. [1982], *Życie i działalność naukowa Stanisława Ruziewicza (1889–1941)* (The life and academic contributions of Stanisław Ruziewicz), Kwart. Historii Nauki i Techniki 27, pp. 653–689.

Wagon, S. [1985], *The Banach-Tarski Paradox*, Cambridge Univ. Press.

Walczak, M. [1978], *Szkolnictwo wyższe i nauka polska w latach wojny i okupacji 1939–1945* (Schools and higher learning in Poland during war and occupation 1939–1945), Wydawnictwo Ossolineum.

Warhaftman, S. [1927], *Pierwszy Polski Zjazd Matematyczny* (The First Polish Mathematical Congress), Mathesis Polska 2.

Wiener, N. [1958], *I Am a Mathematician*, New York: Doubleday.

Więsław, W. (ed.) [1997], *Problemy Hilberta w pięćdziesięciolecie śmierci ich twórcy* (Hilbert's problems on the fiftieth anniversary of their originator's death), Warszawa: Instytut Historii Nauki PAN.

Więsław, W. [2004], *Listy Wacława Sierpińskiego do Stanisława Ruziewicza* (Letters from Wacław Sierpiński to Stanisław Ruziewicz), Wiadom. Mat. 40, pp. 139–167.

Woleński, J. [1985], *Filozoficzna szkoła lwowsko–warszawska* (The Lvov–Warsaw philosophical school), Warszawa: Państwowe Wydawnictwo Naukowe.

Woleński, J. [1988], *Logic and Philosophy in the Lvov–Warsaw School*, Synthèse Library, Dordrecht-Boston-London: Kluwer Academic Publishers.

Woleński, J. [2001], *Powstanie logiki matematycznej w Polsce* (The emergence of mathematical logic in Poland), in: A. Strzałkowski (ed.), *Recepcja w Polsce nowych kierunków i teorii naukowych* (The reception in Poland of new academic directions and theories), Polska Akademia Umiejętności, Komisja Historii Nauki, Monografie 4, Kraków, pp. 63–85.

Woleński, J. [2011], *Lwowska Szkoła Filozoficzna* (The Lvov School of Philosophy), in: *Universitati Leopoliensi in memoriam*, Materiały sesji dla upamiętnienia trzysta pięćdziesiątej rocznicy jego fundacji, Kraków: PAU, pp. 185–207.

Woyczyński, W. A. [1997], *Szukając Birnbauma* (Seeking Birnbaum), Wiadom. Mat. 33, pp. 137–154.

Woyczyński, W. A. [2001], *Seeking Birnbaum, or nine lives of a mathematician*, Math. Intelligencer 23.2, pp. 36–46.

Wrzesiński, W. (ed.) [1994], *Kaźń profesorów lwowskich* (The execution of the Lvov professors), Wrocław: Wydawnictwo Uniwersytetu Wrocławskiego.

Youshkevich, A. P. [1976], *The concept of function up to the middle of the 19th century*, Arch. Hist. Exact Sci. 16, pp. 37–85.

Zygmund, A. [1938], *Międzynarodowy Kongres Matematyczny w Oslo* (The International Mathematical Congress in Oslo), Mathesis Polska 11.

Zygmund, A. [1981], *Steinhaus and the development of Polish mathematics*, in: [B: Mauldin, 1981; pp. 29–34].

C. Other Mathematical Works Cited

Albiac, F. and Kalton, N. J. [2006], *Topics in Banach Space Theory*, Springer.

Alexiewicz, A. [1950–1952], *On sequences of operations I–IV*, Studia Math. 11, pp. 1–30 and 200–236; Studia Math. 12, pp. 84–92 and 93–101.

Berman, G. [1977], *Frequently cited publications in pure graph theory*, J. Graph Theory 1, pp. 175–180.

Bessaga, C. and Pełczyński, A. [1959], *Własność baz w przestrzeniach typu B_0* (A property of bases in type B_0 spaces), Prace Mat. 3, pp. 123–142.

Bessaga, C. and Pełczyński, A. [1975], *Selected Topics in Infinite-dimensional Topology*, Monografie Matematyczne, Warszawa.

Borel, E. [1895], *Sur quelques points de la théorie des fonctions*, Ann. École Normale (3) 12, pp. 3–55.

Borowiecki, M., Kennedy, J. W., and Sysło, M. M. (eds.) [1983], *Graph Theory*, Proc. Conf. in Łagów, 1981, Lecture Notes in Mathematics 1018, Springer.

Bourbaki, N. [1953], *Espaces vectorielles topologiques*, Paris.

Brams, S. and Taylor, A. [1996], *Fair Division: From Cake-Cutting to Dispute Resolution*, Cambridge: Cambridge Univ. Press.

Cantor, G. [1932], *Gesammelte Abhandlungen mathematischen und philosophischen Inhalts*, Berlin: Springer (revised editions: 1962, 1980).

Dineen, S. [1981], *Complex Analysis in Locally Convex Spaces*, Amsterdam: North-Holland.

Dineen, S. [1999], *Complex Analysis on Infinite Dimensional Spaces*, London: Springer.

Duda, R. [1986], *Wprowadzenie do topologii, Część I: Topologia ogólna* (Introduction to topology, Part I: General Topology.), Biblioteka Matematyczna 61, Warszawa: Państwowe Wydawnictwo Naukowe.

Dunford, N. and Schwartz, J. T. [1958], *Linear Operators, Part I: General Theory*, New York: Interscience Publishers.

Fichtenholz, G. [1926], *Sur la notion de fermeture des systèmes de fonctions*, Rend. Circ. Mat. Palermo 50, pp. 385–398.

Fichtenholz, G. and Kantorovitch, L. V. [1934], *Sur les opérations linéaires dans l'espace des fonctions bornées*, Studia Math. 5, pp. 69–98.

Fréchet, M. [1906], *Sur quelques points du Calcul Fonctionnel*, Rend. Circ. Mat. Palermo 22, pp. 1–74.

Fréchet, M. [1907], C. R. Math. Acad. Sci. Paris, CR 444, pp. 1414–1416.

Fréchet, M. [1929], *L'analyse générale et les Espaces Abstraits*, Atti del Congresso Internazionale dei Matematici, Bologna, 3–10 Settembre 1928, Bologna: Nicolo Zanichelli, Vol. I, pp. 267–274.

Fusco, N and Pratelli A [2011], *On a conjecture of Auerbach*, Jour. Euro. Math. Soc. (JEMS) 13, No. 6, pp. 1633-1676.

Gelfand, I. M. [1941], *Normierte Ringe*, Mat. Sb. 9, pp. 41–48.

Hadamard, J. [1898], *Sur certaines applications possibles de la théorie des ensembles*, Verhandlungen des Ersten Intern. Mathematiker-Kongresses, Leipzig: B. G. Teubner, p. 202.

Hadamard, J. [1929], *Le développement et le rôle scientifique du Calcul Fonctionnel*, Atti del Congresso Internazionale dei Matematici, Bologna, 3–10 Settembre 1928, Bologna: Nicolo Zanichelli, Vol. I, pp. 143–161.

Hahn, H. [1922], *Über Folgen linearen Operationen*, Monatsh. Math. Phys. 32, pp. 3–88.

Hahn, H. [1927], *Über lineare Gleichungssysteme in linearen Räumen*, J. Reine Angew. Math. 157, pp. 214–217.

Hausdorff, F. [1914], *Grundzüge der Mengenlehre*, Leipzig: Verlag von Vert; reprint: Chelsea and [C: Hausdorff, 2002].

Hausdorff, F. [2002], *Gesammelte Werke*, Vol. II, Berlin-Heidelberg-New York: Springer.

Hellinger, E. and Toeplitz, O. [1928], *Integralgleichungen und Gleichungen mit unendlich viele Unbekannten*, Enzyklopädie der mathematischen Wissenschaften, II C, 13 (1923–1927), pp. 1335–1602.

Hilbert, D. [1904–1910], *Grundzüge einer allgemeinen Theorie der linearen Integralgleichungen*, I–VI, Nachr. Akad. Wiss. Göttingen, Math.-Phys. Kl., (1904), pp. 49–91; (1905), pp. 213–259; (1905), pp. 307–338; (1906), pp. 157–227; (1906), pp. 439–480; (1910), pp. 355–417.

Hilbert, D. [1912], *Grundzüge einer allgemeinen Theorie der linearen Integralgleichungen* (revised edition: 1924).

Jarnik, V. [1933], *Über die Differenzierbarkeit stetiger Funktionen*, Fund. Math. 21, pp. 48–58.

Jarnik, V. [1934],*Sur les nombres dérivés approximatifs*, Fund. Math. 22, pp. 4–16.

Johnson, J. and Lindenstrauss, M. (eds.) [2001–2003], *Handbook of the Geometry of Banach Spaces*, 2 vols., Amsterdam: North-Holland.

Jordan, C. [1893], *Cours d'analyse de l'École polytechnique*, I edition 1882–1887, II edition 1893–1895.

Khintchine, A. and Kolmogoroff, A. [1925], *Über Konvergenz von Reihen, deren Glieder durch den Zufall bestimmt werden*, Mat. Sbornik 32.4, pp. 668–677.

Kolmogorov, A. [1933], *Grundbegriffe der Wahrscheinlichkeitsrechnung*, Berlin: Springer.

Krasnosielskij, M. and Rutickij, Ya. B. [1961], *Convex Functions and Orlicz Spaces*, Groningen.

Krein, M. and Milman, D. [1940], *On extreme points of regular convex sets*, Studia Math. 9, pp. 133–138.

Kuratowski, C. [1952], *Topologie II*, Monografie Matematyczne 21, Warszawa.

Lebesgue, H. [1902], *Intégrale, longeur, aire*, Ann. Mat. Pura Appl. (3) 7, pp. 231–359.

Lebesgue, H. [1904], *Leçons sur l'intégration.*

Ledoux, M. and Talagrand, M. [1991], *Probability and Banach Spaces: Isoperimetry and Processes*, Berlin: Springer.

Lelong, P. [1971], *Théorème de Banach-Steinhaus pour les polynômes; applications entières d'espaces vectoriels topologiques complexes*, Séminaire P. Lelong, 1970, pp. 87–112, Lecture Notes in Math. 205, Berlin: Springer.

Lévy, P. [1922], *Leçons d'analyse fonctionnelle*, Paris: Gauthier-Villars.

Lindenstrauss, J. and Tzafriri, L. [1977], *Classical Banach Spaces, I. Sequence Spaces*, Berlin: Springer.

Lindenstrauss, J. and Tzafriri, L. [1979], *Classical Banach Spaces, II. Function Spaces*, Berlin: Springer.

Lindenstrauss, J. and Tzafriri, L. [1996], *Classical Banach Spaces I and II; Sequence Spaces, Function Spaces*, Classics in Mathematics (reprint [1977] and [1979]), Berlin: Springer.

Lusin, N. [1930], *Leçons sur les ensembles analytiques et leurs applications*, Paris: Gauthier-Villars; reprint: Chelsea, 1972.

Mazur, S. [1963], *Computable analysis*, Rozprawy Matematyczne 33, pp. 1–111.

Mazurkiewicz, S. [1931], *Sur les fonctions non dérivables*, Studia Math. 3, pp. 92–94; reprint [C: Mazurkiewicz, 1969; pp. 248–250].

Mazurkiewicz, S. [1969], *Travaux de topologie et ses applications*, eds. K. Borsuk, R. Engelking, B. Knaster, K. Kuratowski, J. Łoś, and R. Sikorski, Warszawa: PWN – Éditions Scientifiques de Pologne.

Menger, K. [1932], *Kurventheorie*, Leipzig: Teubner.

von Neumann, J. [1932], *Mathematische Grundlagen der Quantenmechanik*, Berlin: Springer.

von Neumann, J. [1935], *On complete topological spaces*, Trans. Amer. Math. Soc. 37, pp. 1–20.

Orlicz, W. [1992], *Linear Functional Analysis*(the first edition appeared in 1961 in Chinese).

Peano, G. [1890], *Sur une courbe, qui remplit toute une aire plane*, Math. Ann. 36, pp. 157–180.

Pełczyński, A. and Bessaga, C. [1979], *Some aspects in the present theory of Banach spaces*, in: [A: Banach, 1979; pp. 221–302].

Poincaré, H. [1895], *Analysis situs*, Journal de l'École polytechnique 1, pp. 1–121.

Riesz, F. [1910], *Untersuchungen über Systeme integrierbarer Funktionen*, Math. Ann. 69, pp. 441–497; reprint in: [C: Riesz, 1960; pp. 441–489].

Riesz, R. [1960], *Oeuvres complètes/Gesammelte Werke*, Vol. I, Budapest: Akadémiai Kiadó.

Saks, S. [1931], *Sur certaines classes de fonctions continues*, Fund. Math. 17, pp. 124–151.

Saks, S. [1932], *On the functions of Besikovitch in the space of continuous functions*, Fund. Math. 19, pp. 211–219.

Saks, S. [1933], *Théorie de l'intégrale*, Monografie Matematyczne 2, Warszawa.

Saks, S. [1937], *Theory of the Integral*, Monografie Matematyczne 7, Warszawa.

Saks, S. and Zygmund, A. [1939], *Funkcje analityczne* (Analytic functions), Monografie Matematyczne 10, Warszawa.

Schwartz, L. [1950–1951], *Théorie des distributions*, 2 vols., Paris.

Sierpiński, W. [1918], *L'axiome de M. Zermelo et son rôle dans la théorie des ensembles et l'analyse*, Bulletin de l'Académie des Sciences de Cracovie, Classe des Sciences Math., Série A, pp. 97–152.

Sikorski, R. [1953], *On a theorem of Mazur and Orlicz*, Studia Math. 13, pp. 180–182.

Singer, I. [1970], *Bases in Banach Spaces I*, Berlin-Heidelberg-New York: Springer.

Snyder, A. K. and Wilansky, A. [1980], *The Mazur-Orlicz bounded consistency theorem*, Proc. Amer. Math. Soc. 80, pp. 374–376.

Steinhaus, H. [1948], *The problem of fair division*, Econometrica 16, pp. 101–104; reprint: [A: Steinhaus, 1985; pp. 567–569].

Stone, M. [1932], *Linear Transformations in Hilbert Spaces and Their Application to Analysis*, New York.

Tarski, A. [1930], *Une contribution à la théorie de mesure*, Fund. Math. 15, pp. 42–50; reprint: [C: Tarski, 1986, Vol. I; pp. 275–285].

Tarski, A. [1986], *Collected Papers*, edited by S. R. Givant and R. N. McKenzie, 4 vols., Basel-Boston-Stuttgart: Birkhäuser.

Turing, A. [1936], *On computable numbers, with an application to the Entscheidungsproblem*, Proc. London Math. Soc. 42, pp. 230–265, and 43, pp. 544–546; reprint in: M. Davis (ed.), *The Undecidable, Basic Papers on Undecidable Propositions, Unsolvable Problems and Computable Functions*, New York: Raven Press, Hewlett, 1960; pp. 111–153.

Tychonoff, A. [1935], *Ein Fixpunktsatz*, Math. Ann. 111, pp. 767–776.

Volterra, V. [1929], *La teoria dei funzionali applicata ai fenomei ereditari*, Atti del Congresso Internazionale dei Matematici, Bologna, 3–10 Settembre 1928, Bologna: Nicolo Zanichelli, pp. 215–232.

Wiener, N. [1921], *On the theory of sets of points in terms of continuous transformations*, C. R. du Congrès International des Mathématiciens (Strasbourg, 1920), Toulouse, 1921, pp. 312–315.

Wiener, N. [1922], *Limit in terms of continuous transformations*, Bull. Soc. Math. France 50, pp. 119–134 (an expanded version of [C: Wiener, 1921]).

Wiener, N. [1923], *A note on a paper of M. Banach*, Fund. Math. 4, pp. 136–143.

Wilansky, A. [1984], *Summability through functional analysis*, Math. Studia 85, Amsterdam: North-Holland.

Zeller, K. and Beckmann, W. [1965], *Theorien der Limitierungsverfahren*, Berlin: Springer.

Zygmund, A. [1930], *Sur la théorie riemannienne de certains systèmes orthogonaux*, Studia Math. 2, pp. 97–170; reprint: [C: Zygmund, 1989, Vol. I; pp. 285–329].

Zygmund, A. [1935], *Trigonometrical Series*, Monografie Matematyczne 5, Warszawa.

Zygmund, A. [1989], *Selected Papers*, edited by A. Hulanicki, P. Wojtaszczyk, and W. Żelazko, Dordrecht-Boston-London: Kluwer Academic Publishers.

Żelazko, W. [1968], *Algebry Banacha* (Banach Algebras), Biblioteka Matematyczna 32, Warszawa: Państwowe Wydawnictwo Naukowe.

Żelazko, W. [1973], *Banach Algebras*, Amsterdam: Elsevier.

List of Illustrations and Acknowledgments

Figure 1. Map by Roman Duda using some contours taken from "Mały atlas historyczny".

Figure 2. Photo of Lvov University from 1898 by Edward Trzemeski, Wikimedia Commons, public domain. Creative Commons Attribution/ShareAlike License.

Figure 3. Grand Plan of Lvov with Suburbs and Neighboring Areas 1910, from the Ryszard Hubisz Collection, courtesy of Ryszard Hubisz from Krakow.

Figure 4. "Old 'Versity" beside St. Nicholas Church. Courtesy of Stanisław Kosiedowski. Wikimedia Commons, GNU Free Documentation License and Creative Commons Attribution-ShareAlike License.

Figure 5. Postcard of "Lvov Polytechnic" National University © Center for Urban History of East Central Europe.

Figure 6. Stanisław Ulam's master's diploma, Library of the Faculty of Mathematics and Computer Science, University of Wrocław.

Figure 7. Convention of the mathematics and physics circle, Lvov, 1930. Wikimedia Commons, public domain.

Figure 8. Title page of the first volume of "Studia Mathematica", Studia Mathematica, Institute of Mathematics, Polish Academy of Sciences.

Figure 9. The inside back cover of the first volume of "Studia Mathematica", Studia Mathematica, Institute of Mathematics, Polish Academy of Sciences.

Figure 10. View of Fredro Square, as seen from St. Nicholas Street, courtesy of Adam Miranowicz.

Figure 11. Inside the Scottish Café. Public domain.

Figure 12. S. Mazur and S. Ulam on a street in Lvov, sometime around 1935. Photo courtesy of Claire Weiner.

Figure 13. The *Scottish Book* in numbers.

Figure 14. Lvov branch of the PTM in numbers.

Figure 15. Ernest Zermelo in Lvov (Zermelo gave a talk at the Lvov branch of the PTM on May 24, 1929). Public domain.

Figure 16. Stefan Banach. Used with permission of John Greczek.

Figure 17. Hugo Steinhaus. Photo courtesy of Lidia Berger.

Figure 18. Stanislaw Mazur. Archives of the Polish Academy of Sciences, Warsaw.

Figure 19. Kazimierz Kuratowski. Wikimedia Commons, public domain.

Figure 20. Stanisław Ulam. Photo courtesy of Claire Weiner.

Figure 21. Juliusz Schauder, taken (copied) from Oeuvres, PWN, Warsaw, 1978, public domain.

Figure 22. Banach's output in numbers.

Figure 23. Monument, in Wrocław, in memory of the professors murdered in Lvov. Photo by Maciej Kosiedowski, Wikimedia Commons, GNU Free Documentation License and Creative Commons Attribution/ShareAlike License.

Figure 24a. Memorial monument on the Wulecki Heights. Courtesy of Stanisław Kosiedowski.

Figure 24b. Plaque at the Wulecki Heights. Courtesy of Stanisław Kosiedowski, public domain.

Figure 25. Memorial plaque in Lvov cathedral. Courtesy of Stanisław Kosiedowski, public domain.

Figure 26. The seventieth anniversary of the murder of the Lvov professors is marked by the unveiling of a new monument on the site of the execution at the Wulecki Heights. Photo courtesy of Ryszard Cach.

Figure 27. Stefan Banach's grave in the Łyczakowski Cemetery. Courtesy of Stanisław Kosiedowski, public domain.

We thank the Polish Mathematical Society for granting permission to use quotations from various issues of Wiadomości Matematyczne.

On the cover: Grand Plan of Lvov with Suburbs and Neighboring Areas 1910, from the Ryszard Hubisz Collection, courtesy of Ryszard Hubisz from Krakow. Photo of Stefan Banach, used with permission of John Greczek. Photo of Mark Kac, courtesy of the Archives of the Mathematisches Forschungsinstitut Oberwolfach. Photo of S. Mazur and S. Ulam, courtesy of Claire Weiner. Photo of Hugo Steinhaus, courtesy of Lidia Berger. Photo of Stanislaw Ulam, courtesy of Claire Weiner.

Index of Names

(bold page number indicates the location of the biography)

Selected Published Titles in This Series

40 **Roman Duda,** Pearls from a Lost City, 2014
39 **Richard Dedekind, Heinrich Weber, and John Stillwell,** Theory of Algebraic Functions of One Variable, 2012
38 **Daniel S. Alexander, Felice Iavernaro, and Alessandro Rosa,** Early Days in Complex Dynamics, 2011
37 **Henri Poincaré and John Stillwell,** Papers on Topology, 2010
36 **Joshua Bowman,** The Scientific Legacy of Poincaré, 2010
35 **William J. Adams,** The Life and Times of the Central Limit Theorem, Second Edition, 2009
34 **Judy Green and Jeanne LaDuke,** Pioneering Women in American Mathematics, 2009
33 **Eckart Menzler-Trott, Craig Smoryński, and Edward Griffor,** Logic's Lost Genius, 2007
32 **Jeremy J. Gray and Karen Hunger Parshall, Editors,** Episodes in the History of Modern Algebra (1800–1950), 2007
31 **Judith R. Goodstein,** The Volterra Chronicles, 2007
30 **Michael Rosen, Editor,** Exposition by Emil Artin: A Selection, 2006
29 **J. L. Berggren and R. S. D. Thomas,** Euclid's *Phaenomena*, 2006
28 **Simon Altmann and Eduardo L. Ortiz, Editors,** Mathematics and Social Utopias in France, 2005
27 **Miklós Rédei, Editor,** John von Neumann: Selected Letters, 2005
26 **B. N. Delone and Robert G. Burns,** The St. Petersburg School of Number Theory, 2005
25 **J. M. Plotkin, Editor,** Hausdorff on Ordered Sets, 2005
24 **Hans Niels Jahnke, Editor,** A History of Analysis, 2003
23 **Karen Hunger Parshall and Adrian C. Rice, Editors,** Mathematics Unbound: The Evolution of an International Mathematical Research Community, 1800–1945, 2002
22 **Bruce C. Berndt and Robert A. Rankin, Editors,** Ramanujan: Essays and Surveys, 2001
21 **Armand Borel,** Essays in the History of Lie Groups and Algebraic Groups, 2001
20 Kolmogorov in Perspective, 2000
19 **Hermann Grassmann,** Extension Theory, 2000
18 **Joe Albree, David C. Arney, and V. Frederick Rickey,** A Station Favorable to the Pursuits of Science: Primary Materials in the History of Mathematics at the United States Military Academy, 2000
17 **Jacques Hadamard and Abe Shenitzer,** Non-Euclidean Geometry in the Theory of Automorphic Functions, 2000
16 **P. G. L. Dirichlet and R. Dedekind,** Lectures on Number Theory, 1999
15 **Charles W. Curtis,** Pioneers of Representation Theory: Frobenius, Burnside, Schur, and Brauer, 1999
14 **Vladimir Maz'ya and Tatyana Shaposhnikova,** Jacques Hadamard, A Universal Mathematician, 1998
13 **Lars Gårding,** Mathematics and Mathematicians, 1998
12 **Walter Rudin,** The Way I Remember It, 1997
11 **June Barrow-Green,** Poincaré and the Three Body Problem, 1997
10 **John Stillwell,** Sources of Hyperbolic Geometry, 1996
9 **Bruce C. Berndt and Robert A. Rankin,** Ramanujan: Letters and Commentary, 1995

For a complete list of titles in this series, visit the
AMS Bookstore at **www.ams.org/bookstore/hmathseries/**.